Praise for

BLIND DESCENT

"Phenomenal story of exploration and science that no one ever gets to hear about. Stunningly fascinating." —JON STEWART, *The Daily Show*

"Heart-stopping and relentlessly gripping. Tabor takes us on an odyssey into unfathomable worlds beneath us, and into the hearts of rare explorers who will do anything to get there first." —ROBERT KURSON, author of *Shadow Divers*

"Harrowing . . . [James Tabor is] a hell of a journalist." —*The Boston Phoenix*

"A story of glory found and prices paid, of the heights and depths we reach when we ignore limits." —*The Christian Science Monitor*

"Tabor's you-are-there style captures the excitement of these expeditions with the immediacy of an Indiana Jones movie, and the ensuing human dramas which unfold—deaths, divorces, liaisons and love affairs—are equally compelling. . . . Tabor's many talents culminate in this risk-it-all tale of tragedy and triumph." —*BookPage*

"Thrilling. . . . James Tabor examines these fearless pioneers and documents the quest for the deepest place on earth." —Time.com

"Holds the reader to his seat, containing dangers aplenty with deadly falls, killer microbes, sudden burial, asphyxiation, claustrophobia, anxiety, and hallucinations far underneath the ground in a lightless world. Using a pulse-pounding narrative, this is tense real-life adventure pitting two master cavers mirroring the cold war with very uncommonly high stakes." —*Publishers Weekly* (starred review)

"A fascinating and informative introduction to the sport of cave diving, as well as a dramatic portrayal of a significant man-vs.-nature conflict . . . What counts is Tabor's knack for maximizing dramatic potential, while also managing to be informative and attentive to the major personalities associated with the most important cave explorations of the last two decades." —*Kirkus Reviews*

"Tabor brings to gritty and frightening life a little-known and fascinating niche of extreme exploration. . . . A gripping and well-written account of the treacherous world of deep cave exploration . . . best suited to true-adventure fans or any recreational readers seeking a pulse-raising tale of real-life drama and grim determination (best avoided by claustrophobes!)." —*Library Journal*

ALSO BY JAMES M. TABOR

Forever on the Mountain

BLIND DESCENT

RANDOM HOUSE TRADE PAPERBACKS
NEW YORK

BLIND DESCENT

THE QUEST
TO DISCOVER THE
DEEPEST CAVE
ON EARTH

JAMES M. TABOR

2011 Random House Trade Paperback Edition

Copyright © 2010 by James M. Tabor

All rights reserved.

Published in the United States by Random House Trade Paperbacks,
an imprint of The Random House Publishing Group,
a division of Random House, Inc., New York.

Random House Trade Paperbacks and colophon are
trademarks of Random House, Inc.

Originally published in hardcover in the United States by Random House,
an imprint of The Random House Publishing Group,
a division of Random House, Inc., in 2010.

LIBRARY OF CONGRESS CATALOGING-IN-PUBLICATION DATA
Tabor, James M.
Blind descent : the quest to discover the deepest cave on earth /
James M. Tabor
p. cm.
Includes index.
ISBN 978-0-8129-7949-7
eBook ISBN 978-1-58836-994-9
1. Caving—History—21st century. I. Title.
GV200.62.T33 2010
796.52'5—dc22

2009033942

Printed in the United States of America

www.atrandom.com

9 8 7 6 5 4 3 2 1

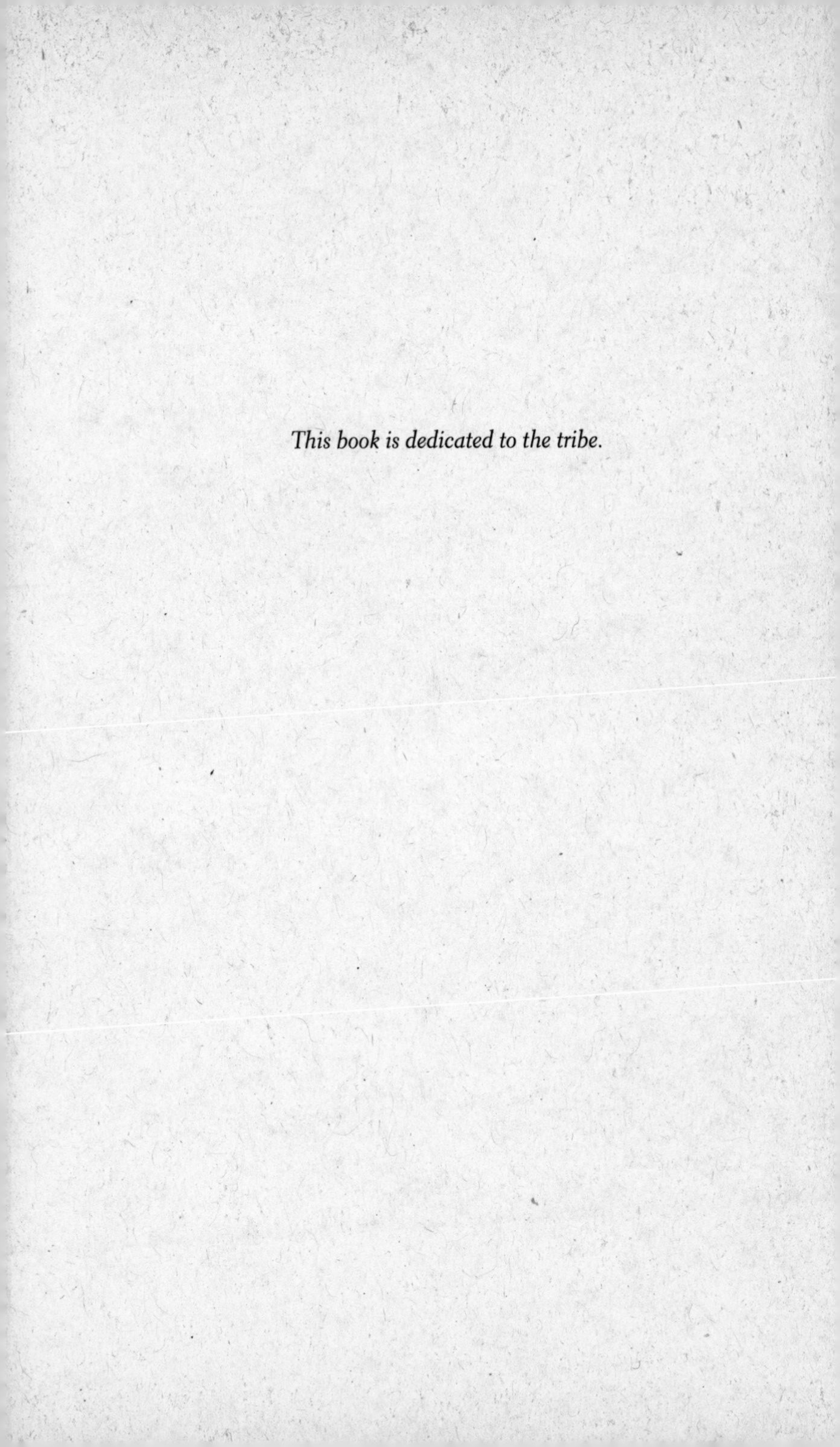

This book is dedicated to the tribe.

Welcome to the world of the deep—
where the strangest things are the people you meet.

—Hazel Barton,
microbiologist and cave explorer

There is nothing more powerful
than this attraction towards an abyss.

—Jules Verne,
Journey to the Center of the Earth

CONTENTS

PROLOGUE

As the fifteenth century began, we believed, absolutely, that the earth was flat.

As the twenty-first century began, we believed with equal certainty that every one of the earth's great discoveries had been made. Almost a century had passed since Peary first trod the North Pole and Amundsen the South. Hillary and Norgay summited Mount Everest in 1953, Piccard and Walsh dove the deepest ocean in 1960. Armstrong and Aldrin walked on the moon in 1969. We played golf and drove a dune buggy there not long after. Surely that tolled discovery's death knell.

But flat-earthers were wrong, and so were those who had prematurely mourned the death of discovery. When the third millennium rolled around, one last great terrestrial discovery *did* still await: the deepest cave on earth. The *supercave*.

Extreme cave exploration is just as exciting, difficult, and deadly as any pioneering feat in mountains, oceans, polar regions, or even off-planet. When he learned about supercaves, Buzz Aldrin said, "I'd thought there could be no environment as hostile as the lunar surface. No more." Thus Aldrin would not be surprised, nor should anyone be, that we stood on top of the world in 1953, but the year 2000 came and went without our having found the bottom of it.

Alien, bizarre, and deadly they most certainly are, but supercaves are not only about adventure. Bill Stone, one of the two great supercave explorers featured in this book, bristled when an interviewer for National Geographic.com asked how he would describe his brand of "adventure."

"Let's first dispense with the adventure label," Stone shot back, adding that "modern, high-tech exploration, which *is* what I do, is quite different. The objective is to advance our knowledge of the frontiers by bringing back new data." Science, in other words, and, indeed, caves are scientific cornucopias as well, furthering research in areas as diverse as pandemic prevention, how the earth was formed, extraterrestrial life's origins, new petroleum reserves, and Mars missions.

And yet, the search for the deepest cave on earth is the greatest epic of discovery and adventure you've never heard of. Despite its drama, danger, and valuable contributions to science, extreme cave exploration remains largely unheralded. In part this is because we prefer our heroes clean and beautiful. Think of our grandest exploration icon, Neil Armstrong: immaculate and pure, his knightly suit burning white against the gray moon and black space. Caving, on the other hand, is by its very nature dirty, dark, and wet.

But there is something else. We've had photographs of mountaineers since the nineteenth century and moving pictures of them almost as long. Good underwater footage from the 1940s exists. And we watched Neil Armstrong actually take his great first step. But for most of its long history, cave exploration remained out of our collective sight and mind. Only quite recently have sophisticated batteries and digital recording technology made it possible to take cameras far down into supercaves, which are thousands of feet deep and many miles long. So while their mountaineering, aquanaut, and astronaut counterparts basked in the limelight, extreme expeditionary cave explorers labored in the dark both beneath the earth's surface and above it.

In fact, the subterranean world remains the greatest geographic unknown on this planet—called "the eighth continent" by some. Mountains, ocean depths, the moon, and even Martian scapes can be—and have been—revealed and explored by humans or our robotic surrogates. Not so caves. They are the sole remaining realm that can be experienced only firsthand, by direct human presence.

By the turn of the twenty-first century, three crucial things had become clear about the last great terrestrial discovery. First, it probably would be made within a decade. Second, it would almost certainly be made in one of two places: the Abkhazia region of the Republic of Georgia or the state of Oaxaca, in southern Mexico. Finally, one of two men would lead the discovery team, which would earn him a place beside the likes of Amundsen and Hillary in the pantheon of exploration. They were the Ukrainian Alexander

Klimchouk and the American Bill Stone, both of whom had devoted their lives to the search for the bottom of the world.

Caves invite juxtapositions of opposites: light and dark, surface and subterranean, safety and terror. Alexander Klimchouk and Bill Stone are both in their fifties, but otherwise they are about as different as men can be, fitting nicely into that list of opposites. Klimchouk is short and slight. Stone is towering and muscular. Klimchouk is quiet, self-effacing, and avuncular. Stone is bold, brash, and commanding. Klimchouk has been happily married to the same woman for decades. Stone divorced in 1992 and has since had a series of relationships with strong, attractive, accomplished outdoorswomen. He is currently engaged to the cave explorer Vickie Siegel, with plans to marry in May 2010. They *are* alike, however, in two key ways: both are scientists and explorers in the classic tradition of Magellan, Amundsen, and Armstrong, willing to risk everything, including their lives and those of others, for the ultimate discovery.

Other explorers and scientists understood the historic nature of such a discovery. They also understood that, for the reasons noted above, it might go relatively unnoticed, which would be doubly tragic. First, because whoever risks everything for such a goal and accomplishes it deserves all the recognition and reward we can bestow. But second, and perhaps more important, this discovery would be not only monumental but sad, marking, as it would, the end of humankind's millennia-old search for the earth's ultimate secrets. So exciting—and perhaps unnerving—was the prospect of this saga's finale that normally staid *National Geographic* magazine had to borrow a phrase from Jules Verne to describe it: "The Race to the Center of the Earth."

As the new millennium dawned, the stage was thus set for an exploration drama unlike any since Roald Amundsen and Robert Falcon Scott competed, head-to-head, with results both horrible and historic, for the South Pole.

This book is the story of the race to make the last great discovery, and of the men and women who won and lost it.

PART ONE
STONE

Rule No. 1: Nothing is impossible (unless it violates the laws of physics).

Rule No. 2: Bend the laws of physics if you can.

—Bill Stone

ONE

STOP.
We have a fatality.

BILL STONE, HALF A MILE DEEP and three miles from the entrance in a Mexican supercave called Cheve, did stop. Red-and-white plastic survey tape hung across the narrow passage he had been ascending. The message, scrawled on notebook paper, was affixed to the tape at chest level, where it could not be missed. Afloat in the cave's absolute darkness, the white paper burned so brightly in the beam of Stone's headlamp that it almost hurt his eyes. The time was shortly before midnight on Friday, March 1, 1991, though that made no particular difference—it was always midnight in a cave.

Stone, a hard-driving man with a doctorate in structural engineering, stood six feet, four inches tall and weighed two hundred hard-muscled pounds. He was one of the leaders (two veteran cavers, Matt Oliphant and Don Coons, were the others) of an expedition trying to make the last great terrestrial discovery by proving that Cheve (pronounced CHAY-vay) was the deepest cave on earth. He had brown hair, a long hatchet face, a strong neck, intense blue eyes, and a prow of a nose angling out between them. Stone was not classically handsome, but it was a striking, unsubtle face men and women alike looked at twice.

Not just now, though. Having been underground for almost a week non-stop, he was gaunt, haggard, and hollow-eyed, his cheeks rough with scraggly beard, and he resembled somewhat the Jesus of popular imagination. A week underground was long, but not extremely so by supercaving standards, where stays of three weeks or more in the vast underground labyrinths were not unusual.

With three companions, he was halfway through the grueling, two-day climb back to the surface from the cave's deepest known point, something like 4,000 vertical feet and 7 miles from the entrance. The note and tape had been strung just before the expedition's Camp 2, where four others were staying. They explained to Stone what had happened. At about 1:30 P.M. that day, a caver from Indiana named Chris Yeager, twenty-five years old, had entered the cave with an older, more experienced man from New York, Peter Haberland. Yeager had been caving for just two years, and going into Cheve was, for him, like a climber who had been on only small Vermont mountains suddenly tackling Everest. This is not a specious comparison. Experts affirm that exploring a supercave such as Cheve is like climbing Mount Everest—in reverse.

Not long after he arrived in camp, more experienced cavers nicknamed Yeager "the Kid." Seriously concerned about the younger man's safety, a veteran, elite cave explorer named Jim Smith sat Yeager down for what should have been a sobering, thirty-minute lecture: don't go into the cave without a guide, carry only a light daypack at first, learn the route in segments, get "acclimatized" to the underground world before going in for a long stay. The warnings fell on deaf ears. Yeager started his first trip with a fifty-five-pound pack, planning on a seven-day stay.

Yeager's problems began soon. Just three hours into the cave, he did not properly secure his rappel rack (a specialized metal device resembling a big paper clip with transverse bars, built for sliding down long, wet ropes with heavy loads in caves) to his climbing harness. As a result, he dropped it. The rappel rack is a critical piece of equipment for extreme cave exploration, probably second in importance only to lights. Without his, Yeager could not continue.

Yeager used his partner's rack to descend to the area where his had landed. Given that a rappel rack is about 18 inches long and Cheve Cave is almost unimaginably vast and complex, this was rather like looking for a needle in a thousand haystacks. Yeager was lucky indeed to find his rack, which allowed him to continue down with Haberland. They did not keep descending for long, however, because they quickly got lost and could not relocate the main route for forty-five minutes.

After seven hours, they arrived at the top of a cliff that had been named the 23-Meter Drop because it was exactly that, a 75-foot free drop from lip to pit that had to be rappelled. By supercaving standards, where free vertical drops hundreds of feet long are common, this was little more than a hop down. Haberland went first, completing an easy rappel without incident. At the bottom he detached his rack from the rope, then moved away to avoid any rocks Yeager might dislodge.

Above, Yeager was wearing standard descent equipment, which included a seat harness similar to those used by rock climbers but beefed up for the heavier demands of caving. A locking carabiner (an aluminum loop, about the size of a pack of cigarettes, with a hinged "gate" on one side) connected the harness to his rappel rack, and the rappel rack connected him to the rope. The rope wove through the rack's bars, like a snake sliding over and under the rungs of a ladder, providing enough resistance for a heavily laden caver like Yeager to control the speed of his descent.

Before going farther, Yeager had to transfer his rack from the rope he had been descending to a new one that would take him to the bottom of the 23-Meter Drop. He made the change successfully, leaned back to begin his rappel, and realized instantly that something was wrong. The rope did not stop his backward-tilting motion. Instead, he kept going, as if tipping over backward in a chair. Somehow his harness had become separated from the rappel rack, which was still attached to the rope.

Instinctively, he lunged to grab the rope and the dangling rappel rack. Had he been carrying no pack, or even a light daypack, it's possible that he might have saved himself by holding on to the rope, or to the anchor bolted to the wall, or perhaps even setting up something called a body rappel. But that would have required almost superhuman strength and would have been extremely difficult even without any load. His fifty-five-pound pack made any such self-arrest impossible, and in another instant he was dropping through space. He fell so quickly that he did not even have time to scream.

Falling rocks can shatter and ricochet like shrapnel; Peter Haberland had moved off and sheltered behind a boulder, so he did not see Yeager land. He realized something was wrong only when he heard a rush of air and the crunching impact of a long fall ending on solid rock. Praying that Yeager had dropped his pack, Haberland called out, but he got no answer.

Within seconds, Haberland found Yeager, lying beside the bottom of the rope. He was in a pool of water three inches deep, on his right side, his face partly in the water, his arms stretching forward, as if reaching for something. Yeager's right leg was broken, the foot rotated grotesquely 90 degrees so that while the body was on its side, the foot pointed up. He had no pulse or respiration, but Haberland turned his face slightly anyway, to keep his mouth and nose clear of the water.

Haberland rushed down to the Cheve expedition's Camp 2, a twenty-minute descent, where he found two other cavers, Peter Bosted and Jim Brown. They left a note hanging from red-and-white survey tape and rushed back up to Yeager's position with a sleeping bag and first aid supplies. When they arrived, they found that some blood had run from his nose, but there were no other changes. All three attempted CPR without success. Chris Yeager was dead.

Understanding precisely why the accident happened requires a detailed knowledge of caving equipment. But the root cause was not equipment failure; it was "pilot error." Yeager entered the cave with too much weight, became fatigued, misused his equipment, and, last and worst, failed to properly secure the locking carabiner that connected his harness to the rappel rack. He apparently made this mistake not just once but twice, the first instance having caused the rack's earlier loss.

LEARNING OF THE ACCIDENT, Bill Stone could only shake his head in dismay. He had been uneasy about Yeager's presence in camp in the first place. Yeager; his girlfriend, Tina Shirk; and another man traveling with them had not been part of the original expedition. After climbing some volcanoes, the three had traveled to the Cheve base camp. Shirk was a competent caver who had been in Cheve the previous year but, with a broken collarbone, was not caving just then. The other man had told Shirk and Yeager that, earlier, he had secured permission for Chris to go into the cave. There is some disagreement about that, but Stone, for one, knew nothing about it. As far as he was concerned, the trio had "crashed" the expedition.

Yeager's death affected everyone. Peter Haberland later wrote in a caving magazine article that he was "shattered at that moment." Tina Shirk was devastated. Other reactions ranged from anger at an overzealous rookie to grief over a young man's death to horror at the reality of a body decomposing down in the cave. For his part, Bill Stone was saddened by the needless loss of a young man's life. He was angered because Yeager's death left the leaders and the team with a thorny problem that could be solved only by endangering others. And he was afraid, not so much of recovering Yeager's body, but that his death might abort the expedition. They could have been on the verge of finding the way into Cheve's deepest recesses and, it was not ridiculous to assume, possibly into history as well. But now it seemed likely that this expedition's time had run out all too soon.

Stone was completely committed to the expedition's mission, financially, emotionally, and physically. The intensity of his work, and his no-nonsense style, left no doubt about that in anyone's mind. He was thirty-nine years old, and if time had not run out for him, he could hear his body clock ticking. Thirty-nine was pushing the upper limit for activities like extreme mountaineering and deep caving, which make such ferocious physical demands on participants.

Like an Olympic athlete who trains for a lifetime to spend minutes chasing gold, Stone knew how precious an opportunity had just been snatched away. It was especially galling to have it stolen by someone who, he believed, had no business being in Cheve in the first place.

Also like an Olympian, Stone was aware that his golden opportunity might never come again in this supercave called Cheve—or anywhere else, for that matter.

TWO

BUT DEATH TRUMPS ALL, and other considerations would have to wait. Yeager—or, rather, his corpse—was now the expedition's responsibility, like it or not. The Mexican authorities, never entirely comfortable with these big cave expeditions, which caused unrest among some insular and superstitious locals, were going to be very unhappy about the death. Worse, they might even want the body, but had none of the skills necessary to retrieve it themselves. That job would fall to Bill Stone, his co-leaders, and the other cavers. The problem was that nobody had recovered a corpse from so deep in a cave like Cheve.

Supercaves present more hazards than any other extreme exploration environment. Just descending into and climbing out of them is exorbitantly dangerous. Recovering a body, dead or alive, from deep within any cave is even worse, increasing that danger by an order of magnitude. The same year Chris Yeager died, a caver named Emily Davis Mobley broke her leg only four hours and several hundred vertical feet from the entrance of a New Mexico cave called Lechuguilla—big but far less hazardous than Cheve. It took more than one hundred rescuers four days to bring her to the surface. One expert estimated that every hour of healthy-caver descent time equaled a day of ascent in rescue mode in Lechuguilla, which was noted for, as cave explorers put it, "extreme verticality."

"Extreme verticality" describes perfectly the part of Cheve through which Yeager's body would have to be hauled. From its entrance, the cave drops like a steep staircase almost 3,000 vertical feet, over a total travel distance of 2.2 miles, before it begins to level off somewhat. It is not one smooth, continuous drop. Those 3,000 feet include innumerable features and formations, with the odd level stretch, but Cheve's main thrust here is *down.* One giant shaft alone is 500 feet deep. Like rock climbers, cavers call such vertical drops "pitches." There are also shorter pitches—*many* of them, in fact—as well as waterfalls, crawl spaces, walking passages, lakes, huge boulder fields, and many more formations, unique and almost impossible to describe except with a camera.

In the entire cave, there are ninety pitches requiring rappels. Thirty-three of those lay between Yeager's body and the surface, including that 500-foot monster. So going back up that way with a body on a litter, at virtually every one of those thirty-three pitches, recovery teams would have to install haul systems of ropes and pulleys and counterweights. The bigger the wall, the more complex the hauling system.

Rigging such haul systems there, particularly on the big walls, would be more dangerous than rappelling down and climbing back up such faces. The work would require that fatigued cavers hang for hours high in the air, in the dark, sometimes under streams of cold water, in painfully biting harnesses, setting bolts and hangers and pulleys. All that would be even before beginning the hauling, which would entail the use of living human bodies as counterweights, among other unpleasant and dangerous tasks. There is more to body recovery, but this gives a hint of its complexity.

Yeager's father, Durbin, arrived several days after the accident with another relative and a caver friend from Indiana. The body, meanwhile, had been secured temporarily not far from the accident site. There followed a week of discussions between the expedition leaders and the Yeager contingent. Stone, not surprisingly, took the lead for his side. He and the others felt strongly that putting expedition members at great risk to retrieve a dead body was unwise. An accomplished climber himself, Stone pointed out that mountaineers often buried fallen comrades in situ. (At the time, something like 130 climbers had died on Everest, and most of those bodies were still up there.) Stone also pointed out, perhaps indelicately but correctly, that recovering the body would be much easier if it were left in the cave for several years, allowing it to desiccate. A smaller team could then more safely retrieve the bones.

Heated discussions followed, particularly between Stone and his co-leaders and Yeager's friend from Indiana. Finally, the law was laid down: no one would be going into Cheve to get that body. In the end, Durbin Yeager understood that a recovery attempt would invite more accidents, and he reluctantly agreed to have his son's body buried in Cheve Cave.

Eleven days after the accident, expedition members carried Chris Yeager's corpse (in what condition can scarcely be imagined) up a short distance to a sandy alcove where an appropriate burial site had been located. They dug a proper grave, interred Yeager with an expedition T-shirt, conducted a Christian burial service, and erected a tombstone with words inscribed in carbide-lamp soot.

The body problem had been solved, but the Mexican authorities remained agitated. Local officials understood that expeditions could make important discoveries, which in turn could stimulate tourism, as had happened in, say, the Central American countries with Aztec and Mayan ruins. The expeditions also contributed cash to local economies when they bought supplies, rented buildings, and hired local porters.

But the cavers also caused unrest among residents, most of whom believed, despite the best efforts of Stone and other leaders, that the gringos were stealing gold and precious artifacts. The locals objected more strenuously to cavers' incursions for religious and spiritual reasons. To them, the caves were home to deities, as sacred as cathedrals and mosques are to Christians and Muslims. The idea of foreigners living in them, defecating and urinating and having sex and leaving garbage, was highly offensive—as those activities would have been in the Vatican or the Grand Mosque of Mecca.

A caver's death was more than enough to upset the apple cart. The cavers knew that the law was different down here. People were sent to jail for any reason, and sometimes for no reason. And there might have been worse places than Mexican prisons, but they were very close to the bottom of the list.

The expedition leaders were ordered to report to a police station in nearby

Cuicatlán. There, the attorney general for the state of Oaxaca grilled Stone long and hard over the telephone. Amazingly, the official demanded that Stone and the others produce Yeager's body, and there were hints about jail if this was not done. Eventually, Stone convinced the man that he could well have *more* bodies on his hands if he insisted on seeing Yeager's. All right, the attorney general growled, but if anybody else dies from now on, a body *will* be produced—or else. This had never been required before. To Stone's way of thinking, it was absurd. It was also, he felt with some resentment, another consequence of Yeager's recklessness.

Surprisingly, the authorities did not evict the team from Cheve or Mexico, and for a brief while Stone thought they had all dodged a bullet. But then a new request to end the expedition arose, and it came from a source as undeniable as the Mexican police, though for different reasons.

The request came not from Oaxaca but from Indiana. Chris Yeager's parents felt that it would be inappropriate to have cavers shambling back and forth over their son's fresh grave in the sandy passage. Cheve was now a burial site; time should be allowed to pass before active exploration resumed.

The expedition abided by the family's wishes, though it meant ending an effort, only recently begun, for which many had sacrificed time and money and had already put themselves repeatedly at great personal risk. In truth, had it been up to Bill Stone alone, the expedition would have continued. Aware of that, some were appalled. How could you keep going—it was just a *cave*, after all—with the body of a freshly killed young man in there, and on your conscience as well?

Stone operated in a different frame. He liked to point out that ships coming to the New World routinely lost 30 percent or more of their crews. Nor, he said, had deaths ever stopped explorers like Scott, Amundsen, or Lewis and Clark. Turning to more recent efforts, he also derided—publicly—NASA's timid approach to space exploration. But the decision at Cheve was not his alone to make.

As word of Chris Yeager's death and its aftermath circulated, it caused a rift in the caving community. A serious, science-minded minority, familiar with the precedents of exploration history, tended to find the in-cave burial acceptable. A much larger, more casual majority thought it abhorrent. By summer, though, the controversy had cooled, turning the spotlight away from the expedition and Yeager's death. Stone, relieved, felt that the incident was behind him.

But it was not. In early 1992, Yeager's Indiana friend, with Tina Shirk's help, organized an expedition to recover the body. They were fortunate to have assistance from a team of gifted Polish cavers, who brought Yeager's body to the surface in three days. The Poles were very good, but the retrieval was easier than it would have been a year earlier, for the very reason Stone had stated to Durbin Yeager. Decomposition had done its work, and the body, while not just a skeleton, did come out in pieces.

Once again, news of "the infamous Chris Yeager incident," as Stone came to think of it, ignited fresh controversy. Many American cavers, Stone among them, were outraged that an upstart team of foreigners had invaded "their" cave. Others, especially Yeager's friends and family, supported the effort.

The fact that two other expedition leaders and Chris Yeager's father were involved in the original decision to leave the body in place seemed to get lost along the way. Partly that was because the outspoken Stone's larger-than-life ways and brusque manner helped make him a natural lightning rod for criticism. Several magazine journalists who spent relatively brief periods with Stone found him less than ingratiating. Their articles in widely read, influential publications such as *Outside*, *National Geographic Adventure*, and *The Washington Post Magazine* reflected that, describing him variously as "domineering," "obsessed," and "pompous."

Stone's hard-driving, type-A way of going alienated some in the caving community as well. Two of the cavers interviewed early in the research for this book had identical responses when Stone's name came up: "He's an asshole." A third echoed that, adding, "And people die on his expeditions."

But it is important to note that the majority of people who have actually gone down into the earth with Stone praise his courage, intelligence, strength, and especially the indomitable perseverance that, decade after decade, enables him to keep pursuing a goal that, each time he nears it, recedes like a mirage.

Not genetically disposed to niceties, Stone also inherited at least two of the hard personality traits found so often in great achievers, explorers not excepted: he is a classic alpha male, and a type-A personality as well. One especially salient type-A characteristic is extreme impatience driven by a maddening sense of urgency. It's an open question whether such people suffer fools or delays less gladly. For them, everything from mowing the lawn to mounting great expeditions feels like a losing race against time, which always seems to be running out.

Niceness aside, type-A and alpha-male tendencies do confer certain advantages, like the willingness—*need,* some would say—to take on challenges that to the rest of us seem incomprehensible at best and insane at worst. Like spending thirty years pursuing the deepest cave on earth, for example. Well before the twentieth century's end, in fact, knowledgeable sources were drawing comparisons between Bill Stone and the driven, brilliant, death-defying Italian climber Reinhold Messner, unquestionably the greatest mountaineer of all time.

The comparison had merit, but one of its corollaries was less frequently mentioned. True greatness is rarely achieved without collateral damage. Like Messner, Stone pursued Olympian goals with relentless, single-minded passion, and it cost both dearly: marriages, families, lovers, security, friendships, and the lives of friends as well.

Stone curtly and unapologetically denied my request to accompany one of his Mexican supercave expeditions as part of this book's early research. A first meeting with him took months to arrange, partly because of his frenetic schedule and partly because he wasn't overly excited at the prospect of frittering away precious hours with a writer. By the time he finally did submit to an interview, I found it hard not to expect some extraordinary combination of Captain Ahab, Mr. Kurtz, and Spider-Man.

Perhaps, though, that should not have been surprising. What ordinary man, after all, would sacrifice everything for the privilege of going to hell?

THREE

ACTUALLY, FOR AN EXPLORER LIKE STONE, Cheve may be more heavenly than hellish, but by any measure it is an extraordinary cave, and the world has an extraordinary couple from California to thank for its 1986 discovery. In December of that year, the Chernobyl fallout was just ending, Reagan's Iran-contra fallout was just beginning, and while their friends were wrapping Christmas gifts in California, Carol Vesely and Bill Farr were thrashing around remote forest, high in the Sierra Juárez, desperately seeking a super-cave. They were going on a tip given them by another caver, Peter Sprouse, who had made an exhaustive study of topographical maps of the area.

Supercaves, vast geologic monsters miles long and many thousands of feet deep, are to the subterranean world as 8,000-meter peaks are to mountaineering; their exploration requires huge, costly expeditions, multiple subterranean camps, and weeks underground. Actually, supercaves are even rarer than 8,000-meter peaks, of which there are fourteen. In 1986, Vesely and Farr could count on the fingers of one hand caves that were contenders for the title of world's deepest.

Like serious mountain climbers, Vesely and Farr were fit, technically expert, adventurous, and iconoclastic. Their lives revolved around caving. Vesely had become a "professional" substitute teacher because it afforded freedom for her true passions, subterranean exploration and discovery. Farr, a software engineer, negotiated work arrangements that allowed him months of free time. Caving was their true career. That other stuff paid the bills.

Vesely, a petite blonde then twenty-nine, and the wiry, energetic Farr, twenty-six, knew that the ultimate supercave—the deepest one on earth—had yet to be discovered. They also knew that these Mexican mountains were prime supercave territory. This whole region was what geologists call karst—a limestone landscape. That and copious local rainfall created ideal conditions for the formation of giant caves.

Thus this Christmas visit, which had them panting in the thin air at 9,000 feet, twelve miles northeast of the nearest town, a day's car travel to the Gulf of Mexico. At sea level in Oaxaca, the climate was tropical. Up where they were, it was pleasantly cool and clear.

After several hours of hiking through mountain forest, Vesely and Farr came upon a gigantic sinkhole half a mile long and three times as wide. A very welcome sign. Sinkholes are created when water flowing underground erodes soluble subterranean limestone, causing the surface to collapse; big sinkholes foreshadow big caves. They continued on down an old logging road, then began following a stream that dove into the woods and kept flowing downhill. They hoped that the stream would eventually lead them to a cave. They could both sense that something big might be in the offing, and soon they were running, weaving between pines that rose like giant slalom poles as they followed the stream's path.

"Wouldn't it be great to find a really grand Mexican-style cave entrance?" Vesely panted. By this she meant something big enough to drive a 747 through. At that very moment they skidded to a stop at the edge of a cliff. The stream plunged over it, falling to a lovely green meadow 25 feet below. And sure enough, a quarter mile away, above the meadow's far end, stood a really grand, Mexican-style cave entrance.

It looked like a giant black mouth with ragged teeth, several stories high and wide enough to hold two Greyhound buses parked end to end. Both

Vesely and Farr were tempted to sprint across that meadow and plunge right in, but they knew better. This was not like making a new route up a mountain, which climbers could preview with maps, telescopes, and photographs. Remotely operated vehicles (ROVs) gave similar advance looks undersea. Even Armstrong and Aldrin saw pictures of their Sea of Tranquillity landing site before they arrived.

But cave explorers like Vesely and Farr could not see the route and so could not anticipate the dangers, a partial list of which includes drowning, fatal falls, premature burial, asphyxiation, hypothermia, hurricane-force winds, electrocution, earthquake-induced collapses, poison gases, and walls dripping sulfuric or hydrochloric acid. There are also rabid bats, snakes, troglodytic scorpions and spiders, radon, and microbes that cause horrific diseases like histoplasmosis and leishmaniasis. Kitum Cave in Uganda is believed to be the birthplace of that ultragerm the Ebola virus.

Caving hazards related to equipment and techniques include strangulation by one's own vertical gear (primary and secondary ropes, rappel rack and ascender connections, et cetera), rope failure, running out of light, rappelling off the end of a rope, ascenders failing on muddy rope, foot-hang (fully as unpleasant as it sounds), and scores more that, if less common, are no less unpleasant.

One final hazard, so obvious that it's easy to forget, deserves mention: getting lost.

Supercaves create inner dangers as well, warping the mind with claustrophobia, anxiety, insomnia, hallucinations, personality disorders. There is also a particularly insidious derangement unique to caves called The Rapture, which is like a panic attack on meth. It can strike anywhere in a cave, at any time, but usually assaults a caver deep underground.

And, of course, there is one more that, like getting lost, tends to be overlooked because it's omnipresent: absolute, eternal darkness. Darkness so dark, without a single photon of light, that it is the luminal equivalent of absolute zero.

Vesely and Farr knew about all these hazards, and the awareness deterred them not an instant. They trotted across the meadow and, up close, found that the cave mouth was *huge*, even bigger than it had looked from afar. They would measure it, later, at 100 feet wide by 25 feet high, but even that was dwarfed by what they found inside. The Entrance Chamber, as they named it, was 225 feet wide by 100 feet high by 650 feet long, big enough, in other words, to have parked three Boeing 757 jetliners nose to tail, with room left over.

The Entrance Chamber sloped steadily downward for 200 yards at about 30 degrees, the pitch of an expert ski run. Its floor was littered with "breakdown"—jagged boulders that had broken from the ceiling over eons and continued to do so unpredictably. Descending through that maze was like clambering down a mountain of wet, junked cars in the dark.

Some 50 yards inside the cave, a giant gray monolith 30 feet high and about 8 feet in diameter rose at an angle from the cave bottom, resembling a smaller, tilting Washington Monument. They passed it, the light growing dimmer with every step, their single flashlight barely adequate. In their skimpy outfits, T-shirts and jeans, they quickly learned that this cave was *cold*. A cave maintains a steady "body temperature," the average surface temperature of its locale. At a lower elevation in Mexico, this could be in the 70s. Way up here, it was about 47 degrees Fahrenheit.

The cave smelled like mud and wet rock and rotting vegetation. It did not have, just yet, the uniquely alive smell found deeper in wild caves. "Alive" here is used advisedly. Many native peoples believe that caves are sentient, living things. This is not entirely unreasonable, given that caves breathe; have active circulatory, digestive, and excretory systems; can contract diseases and suffer injuries and heal many of both; and are constantly growing—just like any other living body.

VESELY AND FARR COULD NOT KNOW for sure yet, but they might have been the first humans ever to set foot in this place, and the power of that possibility charged each moment with electric anticipation. Working their way farther in, they saw two passages leading on from the chamber's wall to their right. A third passage extended from the chamber's deepest point. There the stream that had led them to this cave disappeared into darkness.

They followed the stream down deeper into the cave, leaving the "Washington Monument" 100 yards behind them, stopping finally at an immense, diamond-shaped portal 20 feet wide by 60 feet high. This was really the "door" to Cheve, down at the bottom of the Entrance Chamber and near the end of the "twilight zone," that part of the cave where external light still penetrated. They had never found a passage that large in a cave with much air movement, but wind was whipping down through this one.

All caves breathe. The diurnal pressure changes from solar heating, as well as larger system-related barometric pressure shifts, account for air movement through caves. Little caves sigh. Big caves blow. Supercaves *roar*, some with hurricane-force winds. The bigger the cave, the bigger the blow. With its gusty breath, this one had just given Vesely and Farr the best Christmas gift either could have imagined: the kiss of depth.

FOUR

VESELY AND FARR WERE THE FIRST to see some of the great cave's most impressive features. The next day, they stumbled onto one of its darkest secrets. The Entrance Chamber, it turned out, had at some ancient time been used for rituals and sacrifices. *Human* sacrifices and, to judge from the small, shockingly white bones, many of the victims had been children. Their skeletons lay beneath a cantilevered slab of stone, the killing altar, thrusting up out of the gloomy Entrance Chamber's swirling mist. Later, the pair would learn that these rituals had been carried out by ancient Cuicatecs, Native Americans who had lived in the region a thousand years before the conquistadors arrived and whose descendants inhabited the region still.

Leaving that site and its remains undisturbed, they returned to the giant, windy portal. Just beyond its opening they found a wall, which they rappelled 25 feet to its bottom. They kept following the stream down the steeply sloping cave floor for about 250 feet, where it disappeared into jumbled piles of rock. They retreated, but returned the next day, their last, and descended the same vertical shaft and kept going, rappelling three more short drops of 25 to 30 feet each. The stream reappeared and continued to flow down one side of the section they were descending. They explored about another half mile of virgin cave, turned back, and called it a trip.

VESELY AND FARR RETURNED to the cave twice in 1987, bringing additional troops both times. Though most people shudder and hug themselves at the word "cave," envisioning horribly claustrophobic crawl spaces, supercaves are characterized more by vast open spaces, many of them vertical. On their second foray, in December, they retraced their original route down the series of short drops, followed the stream until it disappeared into a wall, squeezed through a tight vertical opening, and came to the first of Cheve's many remarkable features. It was a huge room, about 150 feet wide and 250 feet high (the U.S. Capitol dome tops out at 288), with a floor that sloped steeply downward.

Beyond, they encountered more vertical pitches requiring rappels, wending their way through sprawling boulder gardens in the intervening sections and then, thankfully, finding a nice stretch of smooth bedrock. Two more short rappels and they stood at the brink of Cheve's first major drop, a pit 165 feet deep. They named this the Elephant Shaft, because it was big enough to have dropped elephants into. The two were experienced at this kind of thing, but even for them, rappelling into a pit like the Elephant Shaft was serious business, requiring balance, guts, specialized equipment, and the experience to use that equipment. Take any one of those out of the mix and fatal results were likely, as Chris Yeager's death would demonstrate in 1991.

Rappelling—making a controlled slide down a rope—is as essential to cave exploration as ice axes and crampons are to mountaineering. Until the 1920s, cavers clambered down rope ladders or had themselves lowered by teams of hefty assistants. Hand-over-hand descents were practical only for very short drops. Rope ladders were safer but awkward, exhausting, and prohibitively heavy for long descents. Brawny helpers cost good money and might wander off with a comely lass or stupefy themselves with strong drink. In addition, descents of hundreds of feet required the use of substantial hardware—winches, revolving drums, and scaffolding. As the complexity increased, so did the likelihood of failure.

Developed by mountaineers in France after World War I, body rappelling initially involved running rope back beneath the crotch, forward around the left hip, up across the chest, and back over the right shoulder. The technique was hard on the groin; worse, it was easy for a rappeller to separate from the rope, with predictable results. By the 1930s, climbers were using metal devices that fastened them securely to the rope, but caving's huge loads and long, wet, mud-greased ropes required "industrial-strength" rappelling devices for control. An enterprising southern caver named John Cole answered that need, in 1966, with what cavers now call the rappel rack, in which the descent rope wove through the stainless steel (or, now, aluminum or titanium) rack's bars.

Racks helped revolutionize caving, but they were not perfect. There are right and wrong ways to feed the rope through the rappel rack's bars. The wrong way was called "the death rig." When a caver leaned back on a death rig, all the rack's bars popped open, launching the hapless victim into what cavers, with typically black humor, call an "air rappel." The vast majority of air rappels are fatal.

Vesely and Farr tied off one end of a long rope, tossed the other into the void, threaded their racks, and rappelled to the Elephant Shaft's bottom. There the beams of their headlamps revealed a big, roaring stream foaming down over a series of pitches, one short waterfall after another, the rushing so violent that fog hung in the air. Many of the falls collected in plunge pools, turquoise water in bronze-colored basins whose outflow created the next fall. Going deeper still, the cavers rappelled down a face of rock like burnished gold, beside a frothing 100-foot waterfall that filled the air with ghostly wisps of mist.

They descended another 200 yards or so of steep passage, and then the cave played a trick on them. The downclimb ended and the floor suddenly started to rise. Scattered with huge breakdown boulders, the giant ramp continued up for about 100 yards, then leveled briefly before dropping again, more steeply, perhaps at about the angle of a ski jump. This precipitous section, which they would name the Giant's Staircase, was covered with boulders perched so precariously that a light touch could have sent some rampaging down the steep slope.

At the bottom of the Giant's Staircase, they came upon one of those features that seem extraordinary on the surface and, down so deep, stretch the imagination almost to its breaking point. It was a shaft 50 feet in diameter and 500 feet deep. For people who have never stood at the lip of such a pit or rappelled to the bottom of one, those three digits have no visceral impact. Cavers understand this. Talking about big drops in caves, they often refer to how long it takes a tossed rock to hit bottom. It took six seconds for a rock to hit the bottom of this pit, which was long enough for it to have reached terminal velocity, or 124 miles per hour. Counting the seconds out—one thousand one, one thousand two—helps take the measure of such a shaft.

Vesely and Farr were now about 2,000 vertical feet deep and 1.5 miles from Cheve's entrance. It is all too easy to reel off numbers like those and move right on, bringing to mind the famous remark by Joseph Stalin that one death is a tragedy but a million is a statistic. To properly appreciate supercave exploration, it is important not to let one's mind and eyes glaze over at the sight of such numbers. Three miles on a level path—or even a mountain route—in daylight is one thing. Three miles immersed in absolute darkness, drenched by freezing waterfalls, wading neck-deep through frigid lakes, spidering up and down vertical pitches, scrambling over wobbling boulders, and belly-crawling through squeezes so tight you must exhale to escape them, is quite another. And 2,000 vertical feet is two-fifths of a mile. Imagine climbing the stairs of two Empire State Buildings in daylight, dry and unburdened. To get out, Vesely and Farr would have to do it in the dark, soaking wet, heavily loaded, on rope the diameter of a man's index finger.

The only proper name for such a fantastic pit, they felt, had to come from fantasy. They called it, fittingly, Saknussemm's Well, after the fictional cave explorer Arne Saknussemm in Jules Verne's classic novel *Journey to the Center of the Earth.* With limited rope, they were able to descend only half its depth before stopping on a bridge of beautiful, pale rock called flowstone. Looking like milk frozen in midflow, the formation was calcite, the white, crystalline form of limestone. Flowstone is always wet, and therefore slick. Perched delicately there, they peered down into blackness that devoured every lumen of their powerful headlamp beams. There was, quite obviously, much more cave yawning before them. Lacking additional rope, they could only retreat.

Just how does one "retreat" from 2,000 feet deep within the earth? Rock climbing, slow and brutal with the huge loads they carried, was never an option. Cavers like Vesely and Farr needed a way to go right back up the rope they had just slid down, using some swami-style, gravity-defying magic. Ironically, the magic they found came not from swamis but, quite possibly, from cavemen.

IN 1931 AN AUSTRIAN MOUNTAINEER NAMED Dr. Karl Prusik "invented" a knot that slid up a rope but, when weighted, grabbed and did not slide down. Sailors had long used the same configuration, nautically known as a sliding hitch. Exactly when sailors came up with *their* knot—if, in fact, it was original with them—seems to have been lost in the mists of history. But reef knots and granny knots ten thousand years old have been found, still tied and holding, in plant-fiber ropes. If cave dwellers could do that, might they not also have tied sliding hitches?

Dr. Prusik *is* generally credited with first using the sliding hitch for mountaineering. French cavers soon imitated climbers, "walking" back up ropes with Prusik knots. These knots worked, but they slipped on wet, muddy, and icy ropes. Little machines were better. The first mechanical ascender appeared in 1933, and all ascenders still work the same way, sliding freely up but, when weighted, locking securely in place with a toothed, pivoting cam that bites the rope.

Cavers use two mechanical ascenders, assembled in a "sit-stand" configuration, to climb long ropes. One ascender is attached to the seat harness and chest harness. The rope runs through it. Another, also attached to the rope, is held in one or both hands. Hanging from it is a rope with loops for both feet at the bottom. To climb, the caver hangs from the chest ascender, which supports her weight. She raises her feet, which allows her to slide the handheld ascender up the rope. She then stands in the foot loops and the chest ascender slides up the rope, grabbing as soon as she weights it again. Repeated over and over, the motion looks like a frog kicking, and the technique is called "frogging."

So, having abandoned their airy flowstone balcony, Vesely and Farr frogged their way out of Cheve. They were certainly excited and encouraged, but they wore a self-defensive armor of skepticism, as well. Supercaves teach those who explore them any number of things, and skepticism is a big one. The numbers work like this: hundreds of promising leads produce a few dozen explorable passages, which, most often, end in piles of boulders or flooded tunnels or simply blank walls. Once in a great while, an explorable passage will go, and once in an ever-greater while, one will just keep going. But those, the fabulous ones that have no stop in them, are rare.

Even so, both Vesely and Farr were thinking that, just maybe, this cave could be the real deal. For one thing, it was still going. For another, it was located in karst country. For yet another, air was really honking through this cave, and it had to be going somewhere. Finally, the size of features like Saknussemm's Well indicated that this cave had been forming for a very long time. (Water is an irresistible force but not a fast one; it takes eons to create something like Saknussemm's Well.)

Vesely and Farr were not alone in feeling so. The population of world-class expeditionary cavers was, and remains, smaller even than elite mountaineering's. By 1988, word had spread through the tribe, and in March of that year Vesely and Farr led their first true expedition to Cheve, a team of seventeen that included some of America's most stellar cavers.

One stood head and shoulders above the rest—literally and figuratively. His name was Bill Stone. Then thirty-six, Stone had already devoted almost a decade to discovering the world's deepest cave.

FIVE

BILL STONE, IT'S SAFE TO ASSUME, would have succeeded grandly in any endeavor. One of those gene-pool anomalies people admire, sometimes envy, and occasionally fear, he is graced with genius-level intellect, prodigious physical strength, boundless energy, and ambition that keeps everything else revving at redline.

Stone's father, Curt, had been a professional baseball player with the Cincinnati Reds' organization and, before that, a four-sport athlete in high school and college. Had a German psychopath not come to power, it's likely Curt Stone would have gone on to a career in major league baseball. But World War II blew up the world, and military service knocked Curt Stone out of the ballparks. His dreams derailed, he became a salesman rather than a shortstop.

By 1952, he was married and living in Ingomar, Pennsylvania, where his son, Bill, was born. The youngster was drawn at first not to sports but to science, and quite early in childhood. Despite his own athletic background and lack of scientific training, Curt Stone recognized that his son was something of a science prodigy and gave Bill a chemistry set when the boy was in sixth grade. The present was welcome, but Stone quickly outgrew it and was soon ordering his own chemicals and equipment from major scientific supply houses. By tenth grade he had created a sophisticated basement laboratory.

In high school, Stone was a self-described science nerd who earned straight A's, built and launched homemade rockets, and spent hours performing experiments in his basement lab. By the time he was in his teens, it was clear that Stone, tall and strong, had inherited his father's physical gifts, if not his passion for team sports. In fact, Stone's family was somewhat in awe of the big, rangy kid whose curiosity and intellect were a match for his imposing physique. They sometimes compared him to Doc Savage, the pulp fiction hero of the 1930s and '40s who was a scientist, inventor, explorer, researcher, and musician and who, according to his creator, radiated "Christliness." Bill may not have been "Christly," but to his kid sister, Judy, he seemed the best big brother a girl could want, a funny guy, warm and protective, always happy to include her in escapades and adventures.

In addition to size, he inherited his father's competitiveness, but team sports held little interest for him, perhaps because being just a player, rather than the leader, seemed unappealing. (There were team captains, to be sure, but the chances of becoming one were small and, at the high school level, determined mostly by precocity.) Not playing organized sports, and thus earning no varsity letters, made him feel like a failure. That was not an option even then for one so competitive, so he got into a sport that involved shooting—bullets, not basketballs. Joining the rifle team, he could compete as an individual, responsible to and for no one else, and in his junior year became a varsity letterman.

Disinterest in team sports did not mean that the big, restless teenager had no appetite for excitement and adrenaline, and during his junior year something happened that would turn the rest of his life into a classic quest producing exorbitant amounts of both. He attended a slide show presented by two men whose names he never forgot, Dick Schmidt and Al Haar, from the Pittsburgh Grotto chapter of the National Speleological Society. He also never forgot the particular moment when the cave bug bit and would not let go. At one point Schmidt and Haar showed an image of a caver hanging on a rope in a vertical cave shaped like a huge cylinder, hundreds of feet deep. The top of the immense shaft was carpeted with bright green moss. Sunlight poured down into it, illuminating the caver, who hung like a tiny, glowing spider on a golden thread. Beneath him yawned a vast, bottomless darkness. Something about that image struck a spark deep in Bill Stone's soul, and an inner voice shouted, *I want to do that.* It was a desire he would spend the rest of his life fulfilling.

That such an image could create a life-shifting epiphany is curious. It's not so hard to understand why someone could be galvanized by, say, images of a professional athlete at work, or a marine biologist, or a jet pilot. It's more difficult to understand such instant attraction to places perpetually and absolutely dark, cold, wet, riddled with coffin-tight passages and gigantic, gaping abysses. Part of the reason Stone could feel so excited was that the slide of the rappeller showed none of caving's darker side, so to speak. It showed cave exploration at its cleanest, brightest, and most exciting, with that climber floating in golden light. But the rest of that slide show did reveal caving in all its other "glories," and they did nothing to dampen Stone's enthusiasm.

He promptly joined his high school's new caving group, the NASTY (North Allegheny Spelunking and Traveling Young) People's Club. In a nearby quarry, Pittsburgh Grotto cavers introduced Stone and others to caving's most basic and essential technique: rappelling. As beginners, they were taught the body rappel. Advanced cavers were using rappel racks by then, but beginners body-rappelled, and there was still that crotch problem. Showing an early flash of his problem-solving skills, Stone bought coveralls and had his mother sew leather patches in the right places. Then he went out and rappelled every quarry within fifty miles.

STONE SOON GRADUATED TO THE HUGE pits and caverns that have made West Virginia a caving mecca. There, while still a high school junior, he completed a rite of passage, rappelling to the bottom ("dropping," in caver-speak) of 158-foot-deep Hell Hole Cave in the Mountaineer State's Germany Valley. The experience introduced him to a whole new tribe of cavers, those who lived, and sometimes died, for vertical work.

"Pit freaks," they were called, and Stone quickly developed an enthusiasm for both them and their exhilarating, "yo-yo" style of caving. He dropped Hell Hole Cave almost sixty times. What went down had to come back up, of course. For ascending, he and many other cavers then used those inefficient Prusik knots. Then one day in 1969, Stone encountered a group of southern cavers at Hell Hole with vertical rigs the like of which he had never seen:

stainless steel rappel racks, seat and chest harnesses, étriers (stirrups), and mechanical ascenders. Those cavers had come up from TAG—Tennessee, Alabama, and Georgia—the most cave-rich area of the United States. Their vertical techniques and gear were revolutionizing cave exploration. Using such rigs, explorers could get down into, and back up out of, caves that had previously been inaccessible. Because of the new rigs' efficiency, cavers could carry along unprecedented amounts of weight, which eventually would make possible the extended expeditions supercaves require. For Stone, it was an epiphany, akin to someone showing the Wright brothers a Piper Cub. He understood in a flash the potential of this new equipment and what it foretold about the future of cave exploration.

After high school, Stone studied engineering at Rensselaer Polytechnic Institute in Troy, New York. By then tricked out with the latest vertical gear, when he was not in science labs, he was out caving with such stars of the day as Buddy Lane, Richard Schreiber, and the already legendary Marion O. Smith, who as of this writing has explored more caves than anyone else, ever—over five thousand. Smith introduced Stone to places like 586-foot-deep Fantastic Pit in Georgia's Ellison's Cave, the longest in-cave free rappel in the Lower 48. Fantastic is big enough to swallow the Washington Monument. Dropped from its lip, a rock takes eight seconds to hit bottom.

So long were rappels like these that cavers poured water on both rope and rack to prevent overheating produced by friction as the rope, under tremendous stress, ran through the bars. During his first descent of Fantastic, Stone stopped at 200 feet to do just that. But, committing an ultimate rookie gaffe, he dropped his canteen. Other cavers, including his guide and mentor Smith, were already down in the darkness on the pit's floor, almost 400 feet below.

"ROCK!" Stone screamed—the standard warning for any falling object. Looking down, he could see tiny lights flying in all directions. Except for one that, strangely, didn't move at all. Terrified that he had brained a fellow caver, Stone completed the rappel in agony. Reaching the pit's floor, he found Marion Smith holding the bottom of the rappel rope taut; such an assist made it easier for Fantastic first-timers to control their descent rate. The canteen had hit and exploded a few feet away, but Smith had not budged. Mortified, Stone started gushing apologies. Shrugging them off, Smith dismissed the very close call with one word:

"Happens."

Stone was blown away by the elder caver's sangfroid. Life had just shown him an archetype—Hemingway's iconic "grace under pressure" ideal—that he would strive to emulate forever after, that would shape his leadership style, and that would serve him well during future trials and tragedies.

Later that night, Stone and the other newbies' hike back to camp was raucous as they burned off Fantastic adrenaline. After listening for a while, Smith said softly to Stone, "Son, this ain't where it's at."

They had just dropped the biggest hole in the continental United States. What on earth could one-up that? If this isn't where it's at, asked Stone, then where *is* it? Smith let the question linger for a while and finally let fly another one-word zinger:

"*Mexico.*"

SIX

SEEING THOSE SOUTHERNERS WITH THEIR VERTICAL gear had produced an epiphany for Stone. Marion Smith's "Mexico" sparked another. He soon learned that Mexico was home to gargantuan caves that dwarfed even Fantastic Pit. It was also 2,500 miles from RPI and Troy, New York. Five dollars was not small change to most college undergraduates in those days, including Stone and many of his caving friends. Mounting a Mexican caving expedition, like putting together a major mountaineering effort, was exorbitantly expensive in both time and money. It appeared that neither finances nor RPI's demanding academic schedule would allow a Mexican expedition. But showing an early flash of both the ingenuity and the determination that would mark his later exploits, Stone hatched a plan. He approached RPI's Geology Department and offered to photograph and map geological features in Mexico if the department would authorize the work as a legitimate, credit-earning project. The department head bit. Stone and his cohorts still had to scrounge money and family cars, which they did. But for the next three years, they spent twelve weeks a year—six in summer, six in winter—making Mexico's great caves their classrooms.

BILL STONE GRADUATED FROM RPI in 1974 with a B.S. in civil engineering. He spent the next year there earning his master of engineering degree in structural engineering. He could have looked at MIT, Stanford, and Caltech, but serious cave explorers have a way of wrapping their lives around that activity with unusual fervor. So instead of Cambridge or Palo Alto, Stone ended up in Austin, at the University of Texas. U.T. Austin had perfectly respectable doctoral programs, but that was not the only, nor perhaps even the main, reason for Stone's choice. If Boulder, Colorado, with the Flatirons in its backyard, was a prime breeding ground for top climbers, Austin, with its proximity to great caves American as well as Mexican, played a similar role for cavers.

Stone came away from RPI and upstate New York with more than just a degree. During his time there, he met a pretty, sensitive young woman with lustrous dark hair. A Syracuse native, Patricia Ann Wiedeman was pursuing her B.S. in physical therapy at nearby Russell Sage/Albany Medical College. Like Stone, her academic focus was science. Probably more important, Pat enjoyed sports and the great outdoors. She especially liked hiking, backpacking, and climbing—and, once Stone introduced her to it, caving. They fell in love while Stone was at RPI, and their relationship survived the separation involved when he headed off to U.T. Austin and she stayed back in New York State to finish her own degree.

Within greater Austin there was one bull's-eye concentration of cavers, in an enclave of slightly seedy houses with big storage areas and small rents on Kirkwood Road. America's best cave explorers dwelled here, and Bill Stone took right up with them. They called themselves the Kirkwood Cowboys, and many did live something like the fabled cowpunchers of yore, working temporary jobs to save enough money for caving trips. When their supplies ran out, ending their expeditions, they hightailed it back to Austin for days of drudge work and nights of epic parties, not unlike the cowboys' shenanigans in Dodge and Tombstone.

Appearances aside, those Austin cavers were highly skilled and deadly serious. Their goal was the obverse of Hillary and Tenzing's historical ascent to the top of the world: Austin's explorers were determined to find the world's bottom. They had the techniques and tools to match that ambition, and they believed that one of Mexico's caves would take them to the last great prize.

Dropping Hell Hole had opened the first phase of Stone's caving career. A 1976 Mexican expedition launched the second. Stone accompanied another rising superstar, Georgian Jim Smith (no relation to Marion), then arguably the best caver in North America. The year before, Smith, just twenty, had co-led an expedition that broke the world's depth record, about 4,300 feet, in a formidable French cave called Gouffre de la Pierre Saint-Martin. Now Smith and Stone went on a three-week trip to explore a cave named Huautla (WAWT-la) in Mexico's southern state of Oaxaca (Wa-HA-ca).

Huautla, a hard two-day drive from Austin, was Stone's first excursion into a giant Mexican cave and introduced him to things he could not have imagined. Water, every caver's greatest enemy, made the strongest early impression. December was very wet that year. Foaming rivers roared down through Huautla, creating huge waterfalls so powerful that it felt to Stone as though the cave walls were resonating like gigantic drums.

Such a description may puzzle readers with no experience of caves, as well as those who have been in only the comfortably dry touristic versions, with their elevated walkways and fantastic light shows. Water, though, is as much a part of caves as darkness. A big cave like Huautla, viewed in profile, looks something like a tree, with a vast network of tiny branchlets on the surface connecting to larger branches farther down, which themselves come together still deeper in major pits and passages.

To oversimplify somewhat, the dissolving action of slightly acidic water in limestone substrata creates most caves, including all of the giants like Huautla. (There are two other types, one created by sulfuric acid, the other by flowing lava.) It takes more water to carve out bigger caves than small ones, and supercaves like Huautla require the highest volume of all. Deep in such caves, explorers encounter watercourses big enough to satisfy the most avid whitewater kayakers. What's fun on the surface, however, is more likely to be fatal underground. When rainy seasons swell these torrents, they can make sections of the cave impassable, trapping explorers down deep or simply carrying them away.

Beginning in 1965, other teams had explored Huautla to a depth just over 1,000 feet. In 1976, the first major expedition to return to Huautla in eight years was a joint effort between Richard Schreiber and Bill Stone. In all, elements of the expedition spent three weeks exploring Huautla, camping underground for five full days and nights at one point, which in itself was something of a breakthrough. (To put this in perspective, by then climbers had been living in camps on the earth's highest mountains for half a century. Because of its greater challenges, extended subterranean camping was still in its infancy.) Finding a way around a lake previously thought impassible, they got down to 2,600 feet. Given the depth they had reached, and the volume of water flow they were finding, and the magnitude of the cave's features, they started to believe that Huautla could go and go—perhaps to the very bottom of the world.

Pulled by the call of that deep, from then until 1988, Stone led or participated in a dozen Huautla expeditions. Preoccupied with his doctoral work, he was not, however, part of a major 1977 Huautla expedition during which six cavers pushed the envelope further than anyone had by living at almost 1,800 feet for twelve days, using over a ton of technical climbing gear and 3,600 feet of rope. At 2,800 feet, the cavers had to rappel through a tremendous waterfall, equal in volume to ten urban fire hydrants gushing at full force. A quarter-mile farther on, they reached the San Agustín Sump, the subterranean lake that had forestalled all expeditions since its discovery by Jim Smith and Bill Steele earlier that year.

In common usage, a sump is a place, lower than its surroundings, where liquid collects. Think of the typical basement sump, from which a sump pump draws water, or the gooseneck sump in pipes beneath sinks. Sumps are the same kinds of places in caves, writ very large: long, winding, flooded tunnels. In places they are so tight that divers have to stop, doff their tanks, push them through, and then follow, donning their equipment again only when the passage widens sufficiently.

That description really does not do the procedure justice, though, because you remain connected to your air tanks only by the regulator hose and mouthpiece. It takes not much force at all to yank the mouthpiece from between your teeth, and if that happens, in what is almost certain to be zero visibility, your chances of finding it again before you drown are not so good. In other sections, sumps can be bigger than the biggest highway tunnels, hundreds of feet across and thousands of feet long, and these present their own challenges.

Given the mortality rate among divers who would later try to probe them, "terminal sumps" was an accurate double entendre. At the time, though, they were thought impassable. One of the more famous cavers of the day summed it up succinctly: "A sump is God's way of telling you the cave ends there." They named this one for the region above: San Agustín Sump.

In 1979, Stone co-led an expedition into Huautla to "crack the sump"—that is, find a way to get past it. Cavers always try, first, to find a dry way past a sump. If that proves impossible, the last resort is scuba diving through the sump. The 1970s were just a couple of decades removed from scuba diving's Stone Age, but that did not slow cave diving's rapid growth. The combination of clunky equipment, hordes of bright-eyed tyros, and the dearth of formal training programs made the '70s the deadliest decade of all for a notoriously lethal activity. (Cave-diving technology today is sophisticated and computer-based, but it remains scuba's deadliest application.)

Stone took those 1970s techniques and equipment down to the sump, along with two other cave divers, Tommy Shiflett and Steve Zeman. Instead of full-sized air tanks, Stone would dive using two "pony tanks," small side-mounted cylinders that scuba divers normally carry only as emergency backups. Over his wet suit he would wear a climbing harness tied to a heavy rope because, though the current was not too strong where he would enter the water, he feared that the sump might end in a waterfall that would suck him over its lip and drown him deep in the cave. To save weight, he would haul down into the cave neither fins nor a buoyancy compensator (an inflatable vest divers use to float on the surface and maintain neutral buoyancy at depth) nor the lead weights divers wear to counteract the positive buoyancy of their bodies, their wet suits, and the gas in their tanks.

What ultimately transpired illustrated vividly why the '70s were so deadly for cave divers. Stone submerged into San Agustín Sump's murky, 64-degree water (35 degrees below body temperature, it's worth noting) and swam into a descending tunnel. Without weights, his buoyancy pinned him to the top of the tunnel, and without fins, he could not propel himself forward. To move on, he had to flip over and, like one of those ceiling-scurrying demons in horror movies, crawl along upside down.

Stone started with 3,000 psi in his small primary tank. After perhaps fifteen minutes, he was down to 1,900 psi and consuming 100 psi with each breath. At 40 feet deep and looking into a bottomless abyss, he knew it was time to turn around. He yanked three times on the line, the signal for his buddies to pull him back. Nothing happened.

At that depth, Stone was no longer stuck to the ceiling. In fact, he was not buoyant at all. The water pressure had squeezed so much buoyancy out of his wet suit that he was suddenly sinking. His long, soaked line began dragging him down even faster into the bottomless sump. Without a buoyancy compensator, he had no way to stop sinking. Without fins, his bare feet provided almost no propulsion. The faster he sank, the deeper he went, and the deeper he went, the faster he sank. A cave diver's ultimate nightmare, and one from which Stone could not save himself by awakening.

SEVEN

ON THE VERGE OF PANIC AND at the edge of death, Stone actually calmed himself by replaying one of Jim Smith's blackest jokes about cave diving:

"You can't get buried no deeper for no cheaper."

That might not have soothed many people, but it says something about his sangfroid and sense of humor that it did calm Stone. He grabbed a wall and started clawing his way back. By that time, though, he was down to 700 psi and still consuming about 100 psi with each breath. He had his second pony tank hanging from one side, but its regulator mouthpiece was hard to reach and he was afraid that while searching for it he might run out of air in his first tank. In addition, he had stirred up clouds of silt, from the talc-fine powdered rock that forms the bottom of many caves' water bodies. This silt is so fine that, once aroused, it remains suspended in the water for some time; in water without a strong current to wash it away, it can hang there for hours. Silt presents two dangers: reducing visibility and befouling delicate scuba regulators. Better but by no means fail-safe today, regulators were notoriously finicky and failure-prone back then. It could take precious little silt to bollix Stone's backup.

Suddenly the choice was taken out of his hands. The rope jerked and he was being reeled in like a trophy bass hooked by a panicked angler, banging helmet and body and tanks against the jagged rock walls. Earlier, his handlers had not felt Stone's tugs because an outcrop had snagged the rope. Suffering a minor panic of their own, they'd talked for a few minutes, then decided to haul him in pronto. The fast ride out may have saved his life, but could have taken it as easily, had one of those collisions with Huautla's ragged walls stripped the regulator from his mouth. As it was, he surfaced with about 300 psi—three breaths—left in his tank.

Stymied by Huautla's top end, Stone next decided to attack through its bottom, so to speak. Huautla's mouth sat high on the side of a mountain over Santo Domingo Canyon, through which the Santo Domingo River flowed. Huautla's resurgence—the place where all the water draining through it finally poured into the river—could be a cave that would lead *upward* to the main Huautla system. A fine theory, but first the resurgence had to be found.

Reconnoitering in on May 3, 1981, he did just that, and he had a special companion to share the celebration: Pat Wiedeman, who had come with him on that year's expedition. The couple had become ever closer in the years since Stone's move to Texas. Less than three months later, they would get married; it was seemingly a match made in heaven. They loved each other fiercely, and not only did Pat share Bill's fondness for outdoor adventures, she could keep up with him, ascending as well as going down. In the spring of 1982, they would climb Alaska's 20,320-foot Mount McKinley, North America's highest peak and one of the world's toughest mountaineering challenges. Few women made it to the summit back then, but Pat did, and by the tough Muldrow Glacier Route, to boot. She also became a proficient cave diver.

Their 1981 find, a cave called Peña Colorada, opened at the bottom of the mountain on which, higher up, Huautla's mouth gaped. Visualize a gigantic spigot (Peña Colorada Cave) at the bottom of a vast holding tank (the mountain) with a giant funnel on top (Huautla Cave). In the rainy season, enough water gushed straight from Peña Colorada's mouth to create a raging, 60-foot-wide river. In the dry season, though—the time of their visit—the terrain became passable. A 100-foot-deep, steep sand slope dropped to a walkable, descending cave passage, which itself shortly came to a flooded tunnel.

This provided enough confirmation for Stone to pursue his plan of connecting the resurgence with Huautla's deepest known point, six miles uphill and almost 2,000 feet higher in elevation.

Back in 1980, fully committed to deep cave exploration as a parallel profession, Stone had created the United States Deep Caving Team, a nonprofit organization that, according to its website, is "dedicated to the exploration, study, and public awareness of the earth's last remaining frontiers and to the development of the technologies necessary to achieve those aims." The USDCT, with its tax advantages for organizer and sponsors, would become the launch platform for most of Stone's subsequent major expeditions, including his next, the 1984 Peña Colorada effort. It took them two years to pull everything together, but by 1984 he and expedition co-leader Bob Jefferys, another top caver of the day, had assembled a stellar team of divers and trained for two years with the latest high-tech diving gear. They expected that 30 percent—about 2 miles—of the cave's 6-mile length would be flooded. This would require them to stock underground camps and live in them for days, perhaps even weeks, at a time, beyond flooded sumps.

In late February 1984, their Peña Colorada expedition invaded Santo Domingo Canyon. A core team of twelve divers had signed on for four months. This was a massive effort even by Himalayan and supercaving standards. There were over forty sponsors, including Rolex, General Electric, the Explorers Club, and other companies with deep pockets and dreams of better images. It would be a mistake to infer from this surfeit of support that extreme caving had suddenly gone mainstream. It had not, but, like extreme mountaineering, it held the attention of a small group that corporations coveted because its demographics mimicked those of the Explorers Club: educated, successful, affluent people.

Companies like Rolex and General Electric do not come knocking, money bags in hand, and shucking for dollars does not come naturally to the type-A alpha male; but, as expedition leaders from Columbus to Hillary have done, Bill Stone went begging for the better part of two years. A glimpse of the process reveals that for serious exploration, more time is spent in boardrooms than in the wild.

First came the proposal. Bill Stone had to produce not one, but, with over forty sponsors, many versions. What worked for Rolex, for instance, wouldn't necessarily be persuasive to GM or GE. An application to the National Geographic Society for a major grant was typical: thirty pages of dense verbiage and documentation, including biographies of himself and the other expedition participants, lists of all other media contacts and funding applications, microscopic budgets for every dollar requested, and a project justification that ran almost two thousand words—the length of a full magazine article. And that was just one application.

Like most expedition leaders, he was a one-man show. Hard as it may have been, proposal writing was probably the least painful part of the process. The most painful was the making-nice part, which came next.

It may have been less frustrating and humbling than, say, pitching screenplays, but for a proud man like Stone, it was agony. Once in the executive suite, he had to sing for his supper, and do it well, because equally intelligent, intense, driven people had done their act before he got there and others would do so soon after he left. Stone claimed that he would do anything for a sponsorship except smoke, and that left a lot.

Asked a number of times what explorer in all of history he most admired, Stone's answer never varied: *Columbus.* And what about the great navigator so impressed him? Courage? Leadership? Sailing skills? Well sure, but more than anything else, Stone affirmed, it was the wily Genoan's skill at snaring sponsors.

EIGHT

FOR THE 1984 PEÑA COLORADA EFFORT Stone had to seek help not only from corporations; he needed massive logistical and transport help as well, and for that he went to the Mexican government, which put army units to work for the expedition. He also hired local Indians, machete experts all, who hacked a campsite out of raw jungle at the end of a little-used burro trail—the only way in or out. Two hundred porters and sixty-five burros hauled in eight tons of supplies and equipment, including seventy-two scuba tanks and a vast array of diving gear. These were high-tech, ultralight tanks of Stone's own design, weighing two-thirds less than conventional tanks, but they still constituted a huge load of almost fifteen hundred pounds.

"It was a mountain of gear," Stone said at the time. True enough, and every ounce was still more than half a mile downhill from the actual cave entrance. Everything that went into the cave first had to go up that half-mile hill, on the backs of cavers.

The expedition also included Pat Stone. They were settled in suburban

Maryland by then. Bill had landed a job as a structural engineer with the federal government, after somehow wangling a commitment that assured him of at least three months off, without pay, every year, to pursue his other profession: exploration. Pat had taken up cave diving, and her training in physical therapy qualified her to serve as the expedition's medical consultant, which she did on that and subsequent expeditions. Outdoorswoman that she was, Pat liked being there, Stone liked having her, and she got along with the teams, which were always overwhelmingly male. The match made in heaven seemed to be especially blessed.

Things got off to a reasonably good start. The visibility underwater was astonishing—almost 100 feet. Divers progressed quickly, cracking sump after sump. A towering, muscular, blond American named Clark Pitcairn, just twenty-three but exceptionally skilled at both caving and diving, led a number of the sump penetrations. By March 16, they had reached Sump 7, covering 30 percent, about 2 miles, of the distance separating them from Huautla proper. Sump 7's depth and length, though, would make it the most dangerous diving yet.

Just getting to Sump 7 was a mini-expedition in its own right. It required a two-day approach from base camp: one day to reach Camp 1, itself just short of a mile inside the cave, and a long second day, nineteen hours or more, to reach Sump 7 from there. By March 16, team members had made those hauls countless times, resupplying divers on the lead teams with scuba tanks, food, carbide for their lamps—dozens of things, day after day. Some were doing more carrying than diving. Discontent was simmering.

On March 18, Stone and Jefferys left Camp 1 for the nineteen-hour trip to Sump 7. The next day, Stone made two dives to more than 120 feet deep. A great tunnel, 30 feet high by 60 feet wide, extended as far as he could see. For ten days, Stone and three other divers pushed deeper and deeper into Sump 7. The last attempt ended when Clark Pitcairn, 180 feet deep and over 500 feet horizontally into the tunnel, was hit by nitrogen narcosis. Also called "rapture of the deep," this dangerous condition tends to occur on dives deeper than 100 feet, caused by an overload of nitrogen in a diver's system. It can make the victim feel as giddy and intoxicated as if he has just chugged five martinis. As with getting hammered on booze, nitrogen narcosis affects one's judgment. "Narked" divers have been known to hand their regulators to fish; others have stripped off all their gear and simply swum away, convinced that they didn't need it any longer.

Pitcairn, suffering loss of concentration and coordination, dropped his safety reel and line and had to abort. Though potentially deadly, nitrogen narcosis has two saving graces. One is that experienced divers recognize its onset before they become completely incapacitated. The other is that ascent quickly makes it go away. Still, given the fact that his dive was extremely deep (130 feet is the limit for recreational divers) and that he was so far into this cave, he was very lucky to escape. That finished the diving, for the time being, but left a lot more cave. On his last dive, Stone had peered through crystal-clear water at a tantalizing tunnel literally big enough to drive a locomotive through. The tunnel just kept going and going up into the mountain, as far as he could see.

Everyone regrouped at base camp, with six weeks of expedition time remaining. Not surprisingly, Stone wanted to use the team. He had invested heavily with personal funds, as he did on most of his expeditions, and had overcome many obstacles, including the arduous years of wooing sponsors, which had taken time from his family duties. His commitment to this expedition had been total and unswerving, like his commitment to cave exploration in general. From where he sat, six weeks looked like an eternity. To use that time, though, team members would be facing more weeks of Sherpa duty hauling tanks and other supplies in to Camp 2. And if Stone and other top divers did crack the sump, they might be exploring on the other side of it for days, if not weeks. Meanwhile, back at camp, the less fortunate would be sitting on their hands when they were not slapping mosquitoes.

Stone was up for it, willing to do the hard carrying and the most dangerous diving himself to keep going, come what may. Accordingly, he called a team meeting to announce that it was time for the group to load up and go back for another try. But then, to his immense surprise, he discovered that the team was *not* up for it. One expedition member who had taken charge during his absences said bluntly, "Well, I've been talking to all the other people, and maybe not."

It was mutiny.

THE REBELLION WAS CAUSED AS MUCH by failed leadership as by team unrest; Stone and Jefferys acknowledged as much later, both using the word "mutiny" to describe what happened. The upshot was that most of the team pulled up stakes and bolted. Their bugout wasted weeks of expedition time, leaving Stone rankled and confused. Was it possible that he had invited the mutiny by hogging the ball, having too much fun diving while others did the scut work? He didn't think so. He had also tried to spend time at the rear, paying attention to logistics, organizing support and carries. By his own account, Jefferys had spent more time underground than Stone had.

Regardless, Stone felt that he had made a serious mistake. As one of the expedition's leaders, he should have maintained not only his own mission focus but that of the team as well. Absent that, others had formulated their own mission plan, which involved surf, señoritas, and tequila.

Given the immense investment of resources and time, and the huge risks involved, should the 1984 Peña Colorada mission be considered successful because, under a strong and effective leader, it explored almost five miles of cave, covering one-third of the distance to Huautla? Or should it be deemed a failure because, with ample time and resources left, an expedition-ending mutiny occurred?

It was both, and it was also an important, if painful, lesson in Bill Stone's ongoing education as an explorer and leader. Moving forward, his challenge would be to repeat the former and avoid the latter. There was another challenge as well. Word spreads quickly through the caving community, especially stories about expeditions gone bad, whether because of accidents, fatalities, or mutinies. At least some of the 1984 Peña Colorada mutineers came back muttering about the expedition's early and unpleasant end. They may or may not have said that the mutiny was Bill Stone's fault. But Bill Stone was one of the leaders, and already had something of a reputation, and people would draw their own conclusions. A captain may not be the only cause of mutiny, but he is always its ultimate target.

NINE

IF THE 1984 PEÑA COLORADA EXPEDITION made Bill Stone a sadder but wiser leader, it also fostered a scientific breakthrough that would help transform not only cave exploration but all recreation, work, and science that required time underwater.

The truth was that the mutiny didn't really end the expedition; conventional scuba technology, reliant on air-filled tanks, had done it first. So much diving was required just to get to the known bottoms of giant caves that by the time explorers reached virgin territory, they were essentially out of air. The Peña Colorada expedition, the most ambitious dive exploration attempted up to that time, was a case in point. Stone's team started with seventy-two tanks. Many of those were used just getting to and from Sump 7, leaving few for exploring the terra incognita that was this expedition's reason for being. Stone knew that without some radically different diving technology, supercave exploration was finished. The problem was, that technology did not exist.

Actually, it did exist, sort of, but only for navy SEALs and their like, as one of the members of the Peña Colorada expedition explained to Stone toward the end of that effort. He was a pioneering U.S. Navy captain, diver, and medical doctor named John Zumrick. Stationed at the Navy Experimental Diving Unit (NEDU) in Florida, Zumrick shared Stone's passion for cave diving. He also shared Stone's frustration at finding themselves so close, and yet so far, from linking Peña Colorada with Huautla above. At the expedition's end, therefore, Zumrick suggested that Stone abandon traditional scuba gear and look into devices called rebreathers.

To greatly oversimplify, a rebreather uses chemicals to "scrub" carbon dioxide from a diver's exhaled breath, which it recycles over and over, producing dramatically longer dive times. Again to oversimplify, a single standard scuba tank gives about twenty minutes of dive time at 100 feet. A rebreather can provide *hours*.

Rebreathers in one form or another have been around since the 1600s, when a Dutchman named Cornelius Drebbel invented the first navigable submarine and then had to create a crude rebreather system to keep crewmen from suffocating aboard his oar-powered, leather-hulled submersible. (He heated potassium nitrate in a metal container, producing oxygen and also potassium oxide, which absorbed carbon dioxide.) Three centuries saw gradual and minimal improvements to the concept. Primitive rebreathers allowed a small number of sailors to escape from sunken submarines during World Wars I and II. But there being no major market demand for better rebreathers—crew escapes from sunken subs were tragic but few—rebreathers did not get much better.

Talking with Zumrick, Stone saw in a flash that rebreathers were the future of supercave exploration. But in 1984, there were no rebreathers for civilian use. Military units were exorbitantly expensive and were useful for only short, shallow dives, there being little tactical need for greater depth. Cave divers sometimes went very deep (Clark Pitcairn had hit 180 feet in Peña Colorada), and expeditions required dozens of hours of dive time. In addition, because they were for use in open water rather than in "overhead environments," as divers call caves, the military units lacked redundancies Stone believed essential for cave diving. Finally, navy rebreathers, Stone felt certain, were not tough enough to withstand the harsh beating for weeks on end that expeditionary caving would deliver.

It was a challenge for which Stone was perfectly equipped by nature, education, and training, with that Ph.D. in structural engineering and his diving experience. He started investigating rebreathers literally the day he arrived home from Peña Colorada. Thus began a ten-year project that would run parallel to his extreme caving efforts. When he was not in a cave, he was up to his elbows in prototypes in his basement workshop. He rose daily at 5:00 A.M., worked on the rebreather for three hours, went to work, came home, kissed Pat and the kids, and disappeared into the basement again.

He had to overcome obstacles that had frustrated many other attempts. First, the new unit had to survive caving's battering. Second, it had to be small and light. Third, it had to be absolutely fail-safe, because the supercave environment was less forgiving even than deep oceans and outer space. Open-water divers have access to recompression chambers and medical services. Astronauts on space walks can pop back into their vehicles. Miles into a supercave, divers enjoy no such safety nets. And, oh yes, a cave-exploration rebreather had to afford dive times then simply unimaginable.

Stone named his creation FRED, for Failsafe Rebreather for Exploration Diving. FRED's gestation was often agonizing. Weeks sometimes passed without progress. Finally Stone would turn off the phone and go into seclusion. Day after day and for not a few nights, he sat alone in the basement, drawing, doodling, racking his brain, climbing the walls. Eventually came the "Eureka!" moment.

If he was having success slowly but surely downstairs, things were a bit less joyful upstairs. For one thing, he was pouring hundreds of thousands of dollars, some borrowed, some personal, into FRED. For another, during these years, the outside world was not all that he was blocking out. Pat, as well as the kids that started coming along, got less and less of his time. Regardless, he worked like this for three solid years, and on December 3, 1987, he rolled out his first prototype—literally—using a dolly and a friend to move the 205-pound behemoth FRED from his truck to the clear water of Wakulla Springs in Florida.

Size was not FRED's only unique quality. He—sorry, *it*—was unlike any other rebreather, ever. For one thing, FRED was really twins. There were two complete computerized systems, and each could do everything independently. In addition, FRED was the first to use volatile but efficient lithium hydroxide as its scrubber.

Stone had to strap himself to FRED in the water, but he did not sink like a man tied to a boulder because he had designed FRED to be neutrally buoyant. One test diver later said that using FRED was like diving with a Volkswagen on his back. That was an exaggeration, but FRED *was* bigger than a Volkswagen's engine and weighed about the same. In Florida, Stone made a nicely controlled descent to the bottom at about 35 feet, opened a book, and prepared to stay for a while. A long while.

Though he was using one of the best diving dry suits available then, after twelve hours underwater Stone was cold. He asked other divers to bring lead weights, which he carried while running up and down a steep underwater sand slope. Repeating the process periodically to generate warmth, he finished his first book and started another.

After twenty-four hours of this, Stone finally surfaced, setting a new record for underwater time using scuba gear by a wide margin. His creation, two hundred times more efficient than traditional scuba equipment, had scored a historic triumph. Amazingly, he had used only half of FRED's capability for this experiment. With the other half enabled, Stone could have doubled his underwater time. His only real challenges had been staying warm and awake. Near the end, a diver and physician friend named Noel Sloan (who would later play a key role in Stone's tragic 1994 Huautla expedition) helped Stone stay alert by kicking him whenever he nodded off. (Well, there *was* one more challenge: at one point the urine receptacle in the dry suit ruptured, so Stone spent the final hours underwater sloshing around in his own pee.)

At 205 pounds, though, FRED was useless for cave diving. Stone put FRED on a diet, refining and miniaturizing system after system. By 1989, the 105-pound MK-II appeared. It would still have to be carried down into caves in pieces and reassembled for diving. But once in the water, this son of FRED would be the answer to a dream.

As an interesting aside, all this time, the U.S. Navy had been working on its EX-19, a rebreather it hoped would allow dives to 450 feet. Despite having an oceanic budget and unlimited manpower, the swabbies came up dry. During a 1984 test, their unit sprang leaks, disabling the electronics. Rumors circulated about EX-19 divers blacking out underwater. After ten years, having spent more than $10 million, the EX-19 was deep-sixed.

Stone, working alone and on a shoestring budget in his Doc Savage basement laboratory, accomplished what the United States Navy could not. How did he do that? As Albert Einstein proved, everything depends on your point of view. Most people would say it was a combination of incredible drive, ingenuity, unshakable confidence, and great financial risk, and they would be right. Stone himself saw it as motivation powered by excitement, enabled by scientific expertise, and, more than anything, driven by relentless focus on a lifelong mission. He would be right, as well.

By 1989, Pat Stone had gone on her last expedition with Bill. In 1985, she began her professional career as a registered physical therapist with the Shady Grove Center for Sports Medicine, not far from their home in suburban Maryland, a position she would hold for seventeen years, specializing in orthopedic rehabilitation. By 1989, Pat's answer to the question of how Stone made such a stunning technological breakthrough would have been different, because when he was not down in the basement with FRED, or at his government job, or doing test dives in Florida, he was going on cave expeditions, and her answer, while less lauding, would be as accurate as any of the others.

TEN

FIFTEEN MONTHS AFTER VESELY AND FARR first laid eyes on pastoral Llano Cheve, red and blue and yellow tents dotted the lime-green meadow, making it look like calico. Smells of steaming food and broiling meat and boiling coffee mixed with the music of guitars and harmonicas. People wandered from tent to tent, searching out friends long not seen, hugging and shaking hands. Expeditionary cavers, a small and insular clan, become preternaturally convivial at tribal gatherings. It was like a mountaineering expedition base camp, yes, but perhaps more like an original rendezvous, the fabled gathering of mountain men in America of yore.

For all the joy of reunion, this was deadly serious business, experienced, proficient technicians and scientists determined to venture where humans had not gone, ever, in conditions of extreme hazard. Unusually low water levels promised easier passage through sumps, and all the cavers were eager to get started. Before anyone could begin exploring the cave, teams of riggers had to fix ropes on all the vertical sections, from 50-foot "droplets" to 500-foot

Saknussemm's Well. Readers may be familiar with the term "fixed ropes" from mountaineering, where these aids are nice indeed to have. But climbers can, and do, perform major ascents without them. Cavers, on the other hand, *need* fixed ropes, for obvious reasons. Free-downclimbing something like Saknussemm's Well (a not atypical pit in supercaves) was simply not an option.

Fixing ropes in Cheve was not a matter of stringing one long rope down into the depths. To rig Saknussemm's walls, for instance, the riggers had to install fourteen separate fixed ropes. So many individual sections were required to protect the ropes from dangerous, abrasive areas and the cavers from waterfalls. Every rope junction required a rebelay, or routing anchor.

To set a rebelay, a rigger hung in harness hundreds of feet above the cave floor, with frigid water pouring and spraying all around. Using a four-pound hammer and a handheld bit, or sometimes a ten-pound battery-powered drill, the rigger then made a hole three-eighths of an inch in diameter three inches deep into the solid rock. He next blew dust out of the hole with a metal tube, hammered a steel sleeve into it, inserted a threaded bolt through a stainless steel hanger's collar, and screwed the bolt into the sleeve in the hole in the rock wall. Then he had to repeat the whole process, because every rebelay anchor point required two bolt-and-hanger setups for redundancy. Then he and the other riggers had to repeat *that* whole double-hanger process thirteen more times, because Saknussemm's Well required fourteen rebelays.

WITH THE "NYLON HIGHWAY" (AS CAVE explorers call miles-long and complicated rope arrays) ready for traffic, team members made it to the bottom of Saknussemm's Well. The top had been quite a pleasant place by supercave standards, cool and dry, and much of the trip down had provided stunning views of the giant pit's flowstone walls, glowing white and gold in the cavers' headlamps as if illuminated from inside.

At the bottom, though, they felt as if they were standing inside a car wash. About 250 feet down, a waterfall gushed out of Saknussemm's wall, and by the time that water plunged another 250 feet it was moving at more than a hundred miles per hour. Water does not compress (this is why bridge-jump suicides are as messy as their skyscraper counterparts), so when it hit a caver's helmet traveling that fast, it felt like someone was dumping buckets of gravel from on high. Adding to the maelstrom was a strong wind that blasted the spray around with astonishing force. Travel time for a caver with a light load, after the rebelays were in place, was five hours from the entrance to the bottom of Saknussemm's Well. With a typical expeditionary caver's full load of fifty to eighty pounds, it took team members longer, seven to ten hours, depending on skills and fitness.

Beyond Saknussemm's Well, the going got wetter and stayed that way for some time. The in-cave stream, having gathered considerable force, flowed away from the bottom of Saknussemm's Well and down through a series of shelves not unlike salmon ladders, which was what the cavers named them. More drops and waterfalls eventually brought them to a subway tunnel–like passage, half a mile long, that was nearly level. After a few more short drops and climbs, the cavers found the perfect location for Camp 2, a level, sandy area 2,641 feet deep, 3.1 miles and thirty-three rope drops from the entrance. (An earlier location, Camp 1, at about 1,300 feet deep, had been abandoned because it was judged too close to the surface.)

Various teams pressed on until a major sump stopped them at about 3,140 feet deep. As always, the initial strategy at the sump was to find a way to swim through, climb over, pass around, or go beneath it. Diving was still the choice of last resort. In this case, creative climbing led to a bypass high above the sump. They named this airy, exposed route—"sporting," cavers call such places—Skyline Traverse. Beyond it the lead team, Jim Smith and Ed Holladay, entered an impressive 50-foot-wide, 900-foot-long passage with a dropping ceiling, and then nobody heard from them for a long time.

BILL STONE HAD STOPPED FOR A quick breakfast at the old Camp 1. He was eating when he heard Smith and Holladay, hardware clanking, finally climbing back up toward him. Well before arriving, Smith started yelling at the top of his lungs, "*Booooty . . . booooty!*"

"What happened?" Stone asked, when the two appeared.

Smith smirked at his partner, who was barely able to contain his own mirth. Finally Smith gave the real scoop: "We just busted this sucker wide open. We'll crack a kilometer on the next push." It was a bold prediction but, as things turned out, a correct one.

At the Skyline Traverse's end, Smith and Holladay had negotiated a 50-foot vertical pitch, then stepped over a giant, delicately balanced boulder they titled, respectfully, the Widowmaker. After passing around the big sump and dropping several more pitches, they relocated the main flow of water, appreciably more muscular down here. The next passage revealed the Jekyll-

and-Hyde nature of caves, which, after hours of ugly crawls and scary drops, can ambush explorers with stunning beauty. They downclimbed a series of giant wormholes of exquisitely sculpted black-and-orange rock, leading to a series of deep pools that descended like the steps of a giant staircase. The pools were filled with aquamarine water so clear they could see pea-sized pebbles many feet below the surface. They called this area of the cave the Swim Gym.

Toward the end of this expedition, Smith and Holladay followed the big stream that flowed on down beyond the Swim Gym. An impassable wall of boulders stopped their final thirty-three-hour marathon. *Finis*, for the time being. The expedition had pushed Cheve to 3,406 feet deep and almost seven miles in total mapped length. Seven, of course, was only the one-way mileage. It was a fourteen-mile round-trip and the *second* seven were hardest, because they were all up. Much was dead vertical, but not all. Here and there the gradient eased and the ceiling was high enough to obviate the need for crawling or stooping. Such areas were sufficiently rare that they had a special name: "walking passages." Even these were not truly easy, because the cavers, already fatigued from hours of rappelling and downclimbing, were carrying at least thirty-pound packs, wearing suits that were soaked and mud-caked, often wading through water from ankle- to chin-deep, and contending with either strong head- or tailwinds.

So just walking out was bad enough, but most of the trip out was not like that. There were roughly ninety vertical faces steep and long enough that the cavers had had to rappel them on the way in. Every one of those had to be frogged back up on the way out. Some, like Saknussemm's Well, required detaching from one rope at a rebelay point and reattaching to another, a tiring and complicated process that was also, along with diving, one of the most dangerous aspects of supercaving.

Vertical walls were often washed by powerful cascades of freezing water, and even the redirected ropes could not avoid all of them. Bill Stone had climbed on the big walls of Yosemite, and he likened coming up out of Cheve to ascending El Capitan, through a waterfall, at night. The big difference, of course, was that when you topped out on El Cap, you knew it was over. In a cave like Cheve, when you "topped out" by hitting bottom, the worst was yet to come.

ELEVEN

A YEAR PASSED. STONE WORKED ON his rebreather, making slow, steady improvements. Vesely, Farr, and others interested in Cheve waited out Oaxaca's rainy season, during which flooding made the caves lethal. In March 1989, Vesely and Farr co-led another Cheve expedition with a friend and top caver named Don Coons. Bill Stone once again came along as a team member, accompanied by Pat, for whom this would be the last big expedition. It would last six weeks and involve an *initial* group of twenty-three. But even before the first rope was rigged, a new problem arose: local inhabitants suspected that the explorers were pirating gold from the cave.

In the past, villagers had vandalized vehicles, threatened cavers with machetes, stolen equipment—and worse. In 1968, angry Mazatec Indians nearly killed a female caver named Meri Fish. She was about 175 feet above the cave floor, but still far below the mouth of a nearby cave called La Grieta, when machete-wielding locals chopped her rope while others held her horrified husband at bay on the surface. She had just climbed past a small ledge, which, miraculously, arrested her fall after a few feet. Seconds earlier, or later, and she would have fallen to a bad death on the rocks below.

Years had passed without such incidents, so at first the cavers were inclined to dismiss local grumblings. Though he was not the expedition's titular leader, Bill Stone stepped to the fore and defused the situation by rigging up a screen in the center of the closest village and presenting a slide show about Cheve. He passed around copies of a handout he had written in Spanish that explained what the expedition was doing. They were scientists, he said, and explorers, too, whose sole interest was finding out how deep this cave was. Their ultimate goal, he went on, was to locate the deepest cave on earth, and this one was becoming a candidate for that honor. Again and again, Stone reiterated that their work most assuredly did not include smuggling gold out of the cave.

What probably sealed the deal was a personal Cheve tour, guided by Bill Stone, for local elders. They accepted, though perhaps warily, given their people's ancient conception of caves, but in any case not really understanding what they were in for. He did not have to take them very far before, awestruck and intimidated, they thanked him lavishly and said it would be fine to go back now. Stone's diplomacy quieted the locals' fears for the time being.

After a few initial forays, single-day round-trips soon became impossible, so the team quickly occupied Camp 2, deep in Cheve. In a few ways, camping in a supercave was like doing so above ground. But there were far more differences than similarities, and one of the biggest was that in the cave, it was always absolutely dark. The cavers' eyes would never adjust because there was not a single lumen of light to stimulate their rods and cones. Thus every step taken, every knot tied and soup bowl filled and latrine used and map read required the circle of light from a headlamp or flashlight.

Prolonged, absolute darkness has profound effects on the human body and mind. For one thing, it disrupts the normal circadian rhythms. If they do not use clocks and alarms, cavers on extended stays underground find that their sleep-wake cycles elongate. They may work twenty-four hours at a stretch and then sleep almost as long. Absolute dark can also induce auditory and visual hallucinations, and it weakens the immune system.

Scientists have measured all those effects. Others are also powerful but harder to quantify. For example, darkness is to caving as water is to diving and air is to flying, a *medium,* in other words, that does more than any other aspect of the environment to shape your experience. Cave darkness feels like water on a dive or air on a flight, where air is your means of support, invisible but essential. It has weight and presence, life, a character all its own. Water and air will kill you quickly if you violate your special relationship with them. Darkness can kill just as quickly—or, perhaps worse, much more slowly.

It is one thing to experience absolute darkness for an hour or even a few days. It is another to live in it for weeks on end. The great English polar explorer Apsley Cherry-Garrard endured months of darkness in Antarctica during the winter of 1911–12. In such darkness, he later wrote in his classic of exploration, *The Worst Journey in the World,* Nature "loses much of her healing power when she cannot be seen, but only felt."

We have a special relationship with light; sight is by far our dominant sense. Similarly, we have an equally special, but very different, relationship with its absence. We fear and loathe it. We've been afraid of the dark for millions of years. Perhaps it was even our First Fear, one that, to survive, our proto-human brains installed as *instinct.* We have all felt that spark of panic in a strange, dark cellar, or in the house when lights go out, or in a pitch-black parking lot very late at night.

ON ARRIVING IN CAMP 2, EACH person unrolled a foam pad and a sleeping bag. The caver slept here and moved on, but the pad and bag did not, remaining in place to be used over and over, for the duration of the expedition, by others passing through the camp. The navy does something like this on ships to save space and calls it "hot bunking," so cavers call it "hot bagging." (Some bags got hotter than others.) The best spots were level, in soft sand, and close to water that was running but quiet.

Given the continual flows of water and air through them, supercaves are always noisy. In places, some can sound like 747 engines. One abuse banned by the Geneva Conventions is torture by loud, ceaseless noise, which is exactly what occupants endure day and night in the noisiest cave camps. The worst spots of all were in cacophonous places where the roar was bad and the sleeping was worse, the only available spaces being sloping ledges or vertical walls. There, like climbers on Half Dome or El Capitan, cavers had to fold themselves into hammocks hung from bolts in the rock walls. Somewhere between the two locale extremes came breakdown campsites that had no level floor and offered only spaces between or on top of boulders. Sleeping in those could be like folding yourself around a spare tire in the trunk of a car.

Once sleeping claims were staked out, the team set up its kitchen. Cooking was done on ultralight mountaineering stoves that used butane canisters rather than the more common Coleman stove fuel, also called white gas, which was only slightly less volatile than nitroglycerin. White gas explosions were bad in any context, but suffering one *six miles* underground, where the possibility of rescue was nil, would have been catastrophic.

Shaving weight, with the cavers, was something of a religion. Many cut labels off tea bags and handles off toothbrushes, and took the round cardboard tubes out of toilet paper rolls. It was not uncommon for a team to use only one pot, one bowl, and one big spoon. When a meal was ready (some kind of hot cereal for breakfast, dehydrated or freeze-dried glop for dinner), they dined like ancient Vikings, spoon and bowl passing from person to person until both were licked shiny clean.

For elimination, campsites featured latrine pits. Protocol required that both functions be performed only at the latrine, which would be filled in and covered at the expedition's end (or, if filled to the top, covered over and replaced by a new one). But if a caver awakened from deep sleep and happened to be far from the latrine, or if reaching it required walking through a dangerous rock field, relief might be afforded by a plastic "pee bottle" (which could produce very nasty unintended consequences, as Bill Stone discovered later in Cheve) or just a nearby crack. Human nature being what it was, the longer an expedition lasted, the looser "latrine discipline" tended to become, so that by the end of some longer ones, cavers moved around in camp as though tiptoeing through a roomful of snakes. Fungus growing on spilled food called for added caution.

There was yet another important difference between surface and subterranean camping. Above, campers had sunlight by day and light from fires or electric devices at night. In the pre-LED days, incandescent bulbs consumed much more battery power. To conserve their batteries, cavers often turned their lights off unless working or traveling. Thus in camp, many hours might pass in absolute darkness. Cavers spoke to the sound of each other's voices rather than to faces. They "saw" the same false-image flashes and glows we perceive when our eyes are closed, but they could not interrupt the process by opening their eyes. Hours of this can induce anxiety and hallucinations.

Not the most romantic environment imaginable, all things considered, but teammates having sex was not unheard of in expeditionary caving. Literally not unheard, too, in fact, because except in camps that were thunderous with water or wind noise, concealing the sounds of love on the rocks was virtually impossible, given the noisy plastic groundsheets cavers placed beneath their sleeping bags.

FOR ALL THE WORK REQUIRED TO establish it, Camp 2 was just a way station, of course, and not far beyond it loomed their immediate objective: the solid wall of breakdown that had stopped them the previous year. Bill Farr, wiry and tenacious, led a team that discovered a way through the rock pile, unlocking the rest of Cheve. This was far harder than that short phrase, "discovered a way through the rock pile," made it sound.

Crawling into and through such a huge breakdown pile was, as a veteran caver named Dave Phillips once observed, like being an ant in a jar of marbles. Even that description doesn't really do the experience justice, though, because marbles are packed tight in a jar. Boulders in breakdown piles are not always packed tight. They could have been thousands of delicately balanced tons of rubble needing only one good nudge to let loose. Really, working breakdown of that magnitude was like free-soloing in rock climbing. In both, the price of failure was final and absolute. Come off the wall or shove the wrong rock and you die. *Splat. Squish.* At least in climbing, the cracks and flakes and smooth, no-holds wall were visible. In breakdown, any rock could be the one that unlocked the avalanche.

Finding a crevice through which he could feel some airflow, Farr moved enough rocks to squeeze his body farther in. Then, crawling and worming and contorting, he kept forcing his way forward, wriggling through places big enough to admit his body and moving rocks to enlarge too tight squeezes. Eventually, he came out on the other side.

Toward the end of the expedition, Carol Vesely, Bill Stone, and an Aussie caver named Rolf Adams did a thirty-hour push deeper than anyone had gone, only to be blocked by *another* impassable breakdown pile.

"Well, that's it for this trip," Stone said, bitterly disappointed. Seeing no way to go forward, he turned around. Diving through sumps was one thing. Excavating rock, without bulldozers, was another. Cheve, it appeared, was just another vast waste of time.

TWELVE

"NOT SO FAST," CAROL VESELY COUNTERED. She was far from ready to give up on Cheve. As Stone and Adams looked on, astonished, she miraculously transported herself through a crack they thought impassable. It was like watching one of those superhero movies where Plastic Man oozes under a locked door and opens it from the other side. With her guidance, Stone and Adams removed a few key rocks, joined her, and they kept on going. It took a lot to impress Bill Stone, but that did, and not only because the new passage connected Cheve with another important cave in the area called Puente. Vesely never bragged about the feat, never tried to lord it over Stone or other cavers, and he respected her almost as much for that as for the breakthrough itself.

During the next push, a team of four, which included Bill Stone, pressed on for another half mile or so, descending through a tunnel 35 feet wide by almost 70 feet high. After that half mile, though, they encountered another, even bigger breakdown pile, which stretched from wall to wall and floor to ceiling. Daunted, the team retreated, then established Camp 3, at 4,078 vertical feet, 4 linear miles, and ninety rappels from the entrance.

Not Bill Farr, though. He had not come this far to be stopped by a lousy pile of breakdown, no matter how big. This one would turn out to be 160 feet long, but Farr had no way of knowing that when he started worming and squirming his way into it. The thing could have been a mile long, for all he knew. But he also knew that there had to be a way through—there was *always* a way through.

Farr finally dragged himself out on the other side, looking like a beleaguered caterpillar struggling from a tenacious cocoon—face and hands skinned, arms and legs bruised, caving suit ripped. But what he found beyond that breakdown was so fantastic that it erased both pain and fatigue. Only one name fit: Through the Looking Glass. He stood before the largest passage yet in a cave of Brobdingnagian proportions. This one was 175 feet wide, ran on for almost 1,000 feet, narrowing somewhat for the last 200 yards, and was 175 feet high at its largest point.

Such vast space, so deep, challenges the imagination. Down deep in caves, Bizarro World reversals of many kinds exist. An obvious one is that darkness prevails rather than light. Another is that caves turn the relationship between mass and space on its head. Up top, for example, masses *interrupt* space—Mount Everest, the Great Pyramid, an aircraft carrier. Down under, though, it is just the opposite—mass is interrupted *by* space, and particularly by monstrous voids like this one.

Powerful signs—increasing water volume, the gigantic passages, strong wind—suggested that Cheve could take them closer to the center of the earth than humans had ever been in a cave. But caves, like appearances, are notorious for deception. Just a few hundred yards farther on, *another* massive boulder choke blocked the way. Not even Bill Farr's persistence could crack that one. The expedition was finished. Cheve was now 10.1 miles long, in terms of total mapped passages, and 4,078 feet deep, the second-deepest cave in Mexico but still far from the deepest on earth. At that time, a 5,354-foot Austrian cave called Lamprechtsofen held the record.

ANOTHER EXPEDITION THAT INCLUDED STONE, FARR, Vesely, and other world-class cavers returned in 1990. They succeeded in passing the massive boulder choke, bypassed a new sump, and explored a half mile of stream passage descending along a series of steps and pools. Eventually, gathering strength, the stream became a 100-foot waterfall whose name, Nightmare Falls, reflected its character. The cave descended less steeply, the stream flowing down and through a series of pools and drops so lovely the cavers called them Wet Dreams. At that section's end the stream flowed into a deep, cobalt-blue lake and, try as they might, no one could find a way past it. Someone would have to dive that lake, and that would have to wait for another year, which was not altogether a bad thing. Stone's rebreather, still not finished, might be ready in twelve months. If so, it would usher in an entirely new era of cave exploration. But that, he knew, given its current state of development, was a big if.

That same year produced a separate, critical discovery. Expedition member Jim Smith (he of the "cheap and deep" burial joke) poured about fifteen pounds of nontoxic green dye into the stream that flowed into Cheve's gigantic entrance. Two days later, cavers way down at Camp 3 saw the dye flowing through. Six days after that, green dye flowed into the Santo Domingo River, 11 miles distant and more than 8,000 feet below the cave entrance. Don Coons was down there when the dye came through, turning the entire resurgent river chartreuse. When Coons walked back into base camp in Llano Cheve sometime later, Smith called out, "Did you see any evidence of the dye trace?"

Coons pulled out a toothbrush dyed bright green. "Looked like that for three days," he crowed. "That good enough for you?"

Stone was ecstatic when he heard. Because water flowed from Cheve's top to its bottom, there had to be continuing, unblocked passage all the way down to the river. What's more, an eight-day passage was extremely fast for such a long horizontal distance between cave mouth and resurgence. That might have happened because the water was flowing through channels big enough for people, as well. (Or it could have flowed through myriad small ones, but only someone given to the half-empty-glass worldview would interpret it that way.)

Many more people would now allow that yes, Cheve really just might become the deepest cave on earth. For Bill Stone, there had never been much question, but the dye trace *confirmed* that. Now they would have to get humans down where the green dye flowed, but that would not begin until yet another year passed.

THIRTEEN

IN FEBRUARY 1991, STONE LED HIS first expedition down to Cheve. Co-led, actually, with two other elite American cavers, Don Coons and Matt Oliphant, although, as one who was there noted, there were three leaders, but there was really just one Leader. Sadly, the rebreather was still not ready for cave diving, so to penetrate the terminal sump that had prevented all further progress, they once again hauled scuba tanks and gear, lead weights, and 2,000 feet of safety line all the way down to Camp 3.

From there, they carried everything another third of a mile and 470 vertical feet down to the Cheve terminal sump. On March 1, a man named John Schweyen, then one of the best American cave divers, geared up and slid into the 54-degree water that looked clear and turquoise under the cavers' lights. There was almost no flow and very little silt to reduce the visibility, which, for a sump, was excellent: 20 to 30 feet. The sump floor was coarse tan and brown gravel. Schweyen started swimming, playing out safety line, following the sump as it descended at 45 degrees until leveling out at a depth of 75 feet.

Remaining at about that depth, he continued on horizontally for another 300 feet, then came to a porthole-sized opening that he could see beyond but could not pass through. Schweyen returned safely from what had been a relatively benign dive.

Two scuba tanks were left, but during the descent a valve had been knocked open on one, causing it to lose half its air. This demonstrated once again how inadequate standard scuba was for supercaving; they could not dive the sump safely with one tank and a half-full backup. Well before the team reached the surface, though, it became clear that neither Schweyen nor anyone else would dive again, because on their ascent they found the note about Chris Yeager's death, which ended everything for 1991.

IN 1993, TWO EXPEDITIONS, NEITHER LED by Stone, tried pushing Cheve, using every trick in their books to pass the terminal sump, digging and climbing, crawling and swimming. Every trick except diving, that is, because the expeditions were led by Matt Oliphant and a great caver named Nancy Pistole, and neither were cave divers. They tried, instead, to discover a dry way around the sump. They did find about a mile of new passage, but it did not take them beyond the impeding sump.

Still focused on perfecting his rebreather, which he believed was the key to supercave exploration's future, Stone viewed these efforts from afar, not especially troubled by them. The terminal sump, he felt strongly, would not be cracked with conventional scuba. The extensive underwater investigation required to find a passage out of the sizable sump could be done only with a rebreather. And no one else was developing a rebreather specifically for cave diving. In other words, he held the only key that could unlock that sump—it was just a matter of making final adjustments to that key.

But even for all that, he had never lost sight of Huautla. Cheve took over his life from 1988 to 1991, true, but partly because his rebreather was not ready to tackle the San Agustín Sump. The U.S. Navy's debacle aside, had Stone been able to work on his rebreather full-time and without financial constraints, it's entirely possible he could have produced a workable, expedition-quality rebreather in two or three years. But he did not have those luxuries, and so the project stretched on and on.

It was not the only thing being stretched—so was Pat Stone's patience. If Cheve had consumed her husband's life for four years, from 1988 to 1991, as he himself has remarked, the rebreather project had done much the same thing for the four years preceding 1988. It would do so again in the two years after 1991. That made a decade. By 1991, the Stone family included three young sons, and the focus of Pat Stone's life had naturally shifted from caves to kids. It was not unreasonable to expect a similar change in her husband. As the years passed, though, it became increasingly clear that such a change was not in the offing. With each departed year, Pat's hopes for a traditional family life grew dimmer.

Stone's hopes for the rebreather, on the other hand, were steadily brightening. So much so, in fact, that in 1992 he began planning his most ambitious expedition to date, a 1994 assault on Huautla with a million-dollar budget, scores of sponsors, a cast of dozens, and his rebreather as the star attraction. Now into its third generation and called the MK-III, it was not fully expedition-ready in 1992, but Stone felt confident that two more years of testing, followed by intensive diver training, would make it so. He would need that much time, in any case, to line up sponsors, secure Mexican permissions, recruit team members, and attend to the thousand other details such an expedition requires. Bill Stone, finally, was seeing light at the end of his ten-year tunnel.

The rebreather's path to fruition had not been without some frightening setbacks. In late 1989, a diver named Brad Pecel repeatedly tested Stone's second-generation rebreather, the MK-II, in Florida. On one particular dive, Stone came along as Pecel's buddy. Not twenty minutes after entering the water, Stone watched, stunned, as Pecel started convulsing. On the surface, the violent jerks and spasms of convulsions are terrifying. Deep underwater, where they almost always include spitting out the breathing mouthpiece, they reach a whole new level of horror. Stone followed standard scuba diving rescue procedure, putting a backup regulator into Pecel's mouth and bringing him to the surface as quickly as safety allowed.

Pecel, far luckier than most divers who convulse at depth, survived. But the incident revealed something unsettling: redundancy was a two-edged sword. Like FRED, the MK-II had two completely independent systems. Pecel's support team had prepared his rebreather incorrectly, plugging the oxygen display panel into the wrong system—the one not being used actively. The rebreather then delivered improperly mixed breathing gas to Pecel underwater, resulting in an overload of oxygen in his system. Oxygen toxicity, a familiar diving hazard, causes convulsions.

When he understood that the MK-II rebreather's complexity was the root cause of his accident, at least as he viewed it, Pecel would have nothing more to do with the unit and summarily quit. The other divers did not quit, but the incident left them with concerns. Word got around about the high-tech, complicated units, and rumors circulated that a rebreather malfunction had almost killed a diver. That was only part of the story, but it seemed to be the part that stuck in people's minds.

The passage of time might have allayed many of the concerns, but the next decade turned out to be a deadly one for Bill Stone and his teams.

FOURTEEN

IN THE SPRING OF 1992, Stone planned to take the core team of divers that would spearhead his 1993 Huautla expedition down to Florida's Jackson Blue Springs for intensive rebreather training. First, in light of what had happened to Brad Pecel, in February he had the whole group spend two weeks testing and practicing with the rebreathers at simulated depths in a hyperbaric chamber on Long Island. (A hyperbaric chamber is a massive steel vessel that can create, in dry conditions, the extreme pressure divers will experience underwater.) All went well, and in April everyone trekked down to Florida for intensive in-water training.

Testing rebreathers was not the only science to be done at Jackson Blue. At the time, NASA was trying to better understand team dynamics on long space missions. Part of that was determining what personality traits, if any, might be ideal for them. NASA could simulate such missions, of course, locking people away in mocked-up spaceships for months, but the ever-present "pretend factor" made simulated results unsatisfying. Cave expeditions, on the other hand, were real and, other than the fact that they explored inner space rather than outer, were strikingly similar to space missions. As Mars-bound astronauts would, supercavers had to work and live under incredible stress in dangerous, confined spaces, relying on high-tech life-support systems like Stone's rebreathers, all essentially beyond hope of rescue. To glean as much information as possible, a University of Texas psychiatrist, Dean Faulk, spent several days testing and interviewing the team. Two years later, another University of Texas team would do a much more detailed study of the dynamics of one of Stone's teams.

Faulk was not the only outsider recording interviews and scrutinizing operations. *Outside* magazine had assigned the writer Craig Vetter to do a profile of Bill Stone, who invited him to come along and even to do some rebreather diving. Vetter accepted and was on location for about a week of the team's two-month stay at the springs.

As they had in the dry chamber, both divers and rebreathers passed every test in the incredibly clear, aquifer-filtered water of the famous spring. Stone was reassured by the rebreathers' flawless performance, and the divers, for their part, felt much better than they had with the MK-II in 1989. People were packing up and ready to go on Easter Sunday, April 19. A close friend of Stone's, Rolf Adams (the same Aussie who had been with Stone during that marathon push into Cheve), was one of the best dry cavers in the world and a member of the dive team. He had completed both rebreather training sessions without a problem. Though not an experienced cave diver, he also had completed a basic cave-diving certification course during his stay at Jackson Blue.

Adams wanted to explore a few more of Florida's legendary underwater caves before heading all the way back to Australia. The rebreathers were already packed for transport, but he found another diver and the team doctor, Noel Sloan, the buddy who had kicked Bill Stone awake repeatedly during the FRED test, lounging around in the sun. Sloan had his traditional scuba gear still unpacked. Adams asked to borrow it for a dive in Hole in the Wall, a popular nearby cave.

For a doctor, Noel Sloan had a surprisingly mystical dimension. Right then and there he experienced a frightening premonition, as clear and sharp as a strong radio signal: *This man is going to die.* It was so powerful, in fact, that this team physician, an honest and straightforward man liked by everyone who met him, told Adams a bald-faced lie. Uncomfortable talking about the premonition, he demurred on the loan, saying that he planned to use the gear himself for a dive a little later. It was a complete falsehood, made up on the spot.

Undeterred, Adams eventually borrowed some gear. He and team member Jim Smith took a half-mile boat ride. They donned their gear, submerged, and entered the cave through a hole in the underwater wall. They swam down a shaft, called a chimney, to a depth of 80 feet, leveled off, and continued, descending gradually as they went. The visibility was typically excellent, more than 40 feet, though a fuzzy, reddish silt covered the walls and ceiling where their lights shone.

They swam on, following the white guideline on the cave floor, passing through sprawling rooms with shining white limestone walls and varicolored ceilings, several tight squeezes, and some nondescript tunnels. Having traveled in 2,000 feet and used one-third of their air, they turned around and started back. That was a standard cave-diving procedure called the Rule of Thirds. You used one-third of your air going in, one-third coming out, and kept one-third in reserve. Starting their return trip, then, both divers still had two-thirds of their original air supply, much more than enough to regain the cave's entrance.

They had gone back about 1,000 feet and were 100 feet deep in the water when Jim Smith, with the sixth sense veteran divers develop, felt the absence of his buddy. Looking back, he saw Adams pinned to the ceiling of the cave, his buoyancy apparently out of control, fighting to get his backup regulator into his mouth. He did that, settled down, signaled that he was okay, and they went on, Smith still leading. Just seconds later, Adams grabbed Smith from behind and made the signal all divers fear most, pulling a flat hand across his throat as though slitting it with a knife: *I'm out of air!* Barring equipment failure, it did not seem possible, but that was the frantic signal Adams was giving.

Following prescribed cave-diving protocol, Smith took his primary regulator from his mouth, gave it to Adams, and began breathing from his own backup. That can be tricky in clear, warm, open water. In a tight cave environment, with one diver on the verge of panic, it was devilishly difficult. By the time Adams had the backup regulator in his mouth, both had lost control of their buoyancy and dropped to the cave floor, where clouds of silt enveloped them instantly, reducing the visibility to zero. Overcompensating, they rose too quickly, stirring up even more silt.

Through it all, Adams clutched Smith's chest harness with both hands in what likely would have become a literal death grip for the two of them. It appeared that he was having trouble getting air from Smith's regulator, too, which was very strange, because Smith's primary was working perfectly—he had just been using it himself. Adams's mouth opened, the regulator floated out, his grip on Smith loosened, and he disappeared into the fog of silt.

Smith, distraught and by then low on air himself, managed to swim out of the cave, emerging with his air gauge reading close to zero. He got into the little boat and headed back to camp. Noel Sloan heard the outboard motor approaching and *knew.* He knew with such cold certainty that he found Bill Stone and, without preamble, told him that Rolf Adams was dead. In minutes the unnerved Smith arrived to confirm Sloan's premonition.

A legendary Floridian named Sheck Exley, the father of cave diving (and a good friend of Bill Stone's as well), came over to recover Adams's body. Careful examination of his equipment revealed that there was plenty of air in Adams's tanks and that his regulators were working normally. So were both of Smith's.

Jim Smith, as brave and steady as they come, was shattered by the experience and subsequently gave up cave diving altogether. To this day, Adams's death remains something of a mystery. Various explanations have been put forward, including nitrogen narcosis, arterial gas embolism, and simple panic. There are problems with all those theories. Nitrogen narcosis usually occurs at depths greater than 100 feet; it can happen at 75 feet, Adams and Smith's depth at the time of the accident, but is rare at that depth. Arterial gas embolism, the medical term for bubbles in the bloodstream, can damage the heart, lungs, and brain, but normally results from expansion of respiratory gases during rapid ascents. Smith and Adams were swimming along a nearly level passage. It can also be caused by holding one's breath during any ascent, but Adams was too well trained to hold his breath during a dive. In addition, victims of arterial gas embolism do not usually feel "out of air." More common responses are convulsions and almost instant unconsciousness. Sheer panic seems perhaps the most likely explanation of all. The one thing that can be said with absolute certainty is that Stone's rebreather had nothing to do with Adams's death.

In the end, "death by drowning" was the finding, official but not clarifying, as it would also be in a subsequent death even more closely related to Bill Stone's work. Rolf Adams was one of Bill Stone's closest friends, and the death affected him deeply. He put all his expedition plans on hold and flew to Australia, where he delivered a eulogy at Adams's funeral. The young man's father urged Stone not to cancel his 1993 expedition plans—the last thing Rolf would want, he pointed out—but Stone did so anyway.

Returning home, he wandered around for days, distracted, his mission focus lost. Given his response, it was reasonable to expect a different reaction from caving insiders who had earlier accused him of being insensitive to Chris Yeager's death and who had speculated after the 1989 Pecel incident. Unfortunately—that lightning-rod thing, again—it worked the other way. Stone experienced what he called a "barrage" of attacks from within the caving community, some of which publicly (and wrongly) suggested that he had offered up a friend's life on the sacrificial altar of his rebreather.

Outside's article, published in November 1992 under the title "The Deep, Dark Dreams of Bill Stone," did not say that, exactly; nor did it overtly vilify Bill Stone. The article did capture Stone's type-A behavior: "Stone strides everywhere, or jogs, as if whoever designed the diurnal rhythm of the universe didn't put quite enough hours into the mechanism." It hinted that he might have been less than an ideal family man: "After a full day of designing bridges, he would spend a few minutes getting reacquainted with the family before descending again to the workshop." And it did call him "obsessive."

But Vetter, a respected journalist, also described Stone's contribution to supercave exploration, his persistence and stamina and ingenuity, his far-sighted vision, "his dogged willingness to do whatever it takes." He asked Stone if he ever became discouraged, and the laughing reply, "About once a week," gave a nice glimpse of Stone's humanity and humor. It was, overall, a fair and balanced picture of the man.

Vetter was not present when Adams died, but the accident, described in detail, was the most powerful part of the entire piece. It did make clear that Adams was diving on borrowed, conventional gear, that he was not an experienced cave diver, and that panic probably caused his death. Its final paragraph, though, concerned not Rolf Adams but Bill Stone, who had reached the "point that lies on the way to all unexplored places and demands of explorers a toughness of heart that honors the prize over the price—no matter what."

"A toughness of heart that honors the prize over the price—no matter what." What exactly was that supposed to mean? That Stone was admirable for his indomitable will that overcame all obstacles? Or that he was marching to preeminence over the bodies of his comrades? That final paragraph, the most important single grouping of words in the entire article, was like one of those shifting images that appear to be a vase one moment and two women's profiles the next. The more you looked at it, the harder it became to know just *what* it was. Unless, looking at it, you were the one on whose watch a close friend had just died. Then it was as sharp as the point of a spear.

STONE WAS NOT, REGARDLESS, A MAN to be undone by tragedy or criticism, no matter how painful. Eventually, the Huautla mission started coming into focus again, and the expedition planning got back on track. Sadly, not so his marriage. Pat Stone had viewed the Florida tragedy from afar, but with considerable alarm. She had been in big caves and knew the many ways they could kill you, and she knew, as well, that cave diving increased the odds of a bad death astronomically. Now that she was no longer at his side on expeditions—had not been for several years, in fact—Bill's absences and adventures were exhausting rather than exhilarating. A terrible death, such as Adams's, only added to her stress.

For more than ten years Pat Stone had been standing by her man—and their house, and the kids, and her job—seeing her time with Bill shrink and the debts grow and knowing, despite her hopes, that the clock on their marriage was running down. Rolf Adams's death was the final straw. Soon after Bill's return from Australia, Pat let it all out. In her quiet, firm way—Pat was not a screamer—she said that he was wasting his life, abandoning her, and neglecting their children, all for some godforsaken hole in the ground in Mexico. Then she told him she wanted a divorce.

FIFTEEN

TWO YEARS LATER, AT ABOUT MIDNIGHT on Sunday, March 27, 1994, Bill Stone was back underground, enjoying the world's deepest sleep—literally—in Huautla Cave's Camp 3. His long-awaited Huautla expedition had finally come to pass, and the team had been working in the cave for about six weeks already. Camp 3 was a vertical half mile deep and two miles distant from the entrance, a trip that took good, strong cavers two days to make. The cavern in which Camp 3 sat was the Sala Grande de la Sierra Mazateca, or SGSM. It measured about 200 feet wide, 165 feet high, and 650 feet long, with a total area of about 3.5 acres. It could have contained seven typical McMansions and their half-acre lots.

If not a McMansion, the camp was, at least by deep-caving standards, a four-star abode. There were smooth and level sleeping areas and four Olympian "thrones" constructed by the members of a 1981 expedition who had hauled limestone slabs from around the cavern and assembled them into crude chairs. The camp was strewn with blue and red sleeping bags, mountaineering stoves, shiny cooking pots, waterproof Nalgene bottles full of various powdered foods, and piles of technical caving and diving gear.

Shortly after midnight, Stone was awakened by a sound like the rattling of chains. It was Kenny Broad, his climbing hardware clanking as he arrived from Camp 5, which was 550 vertical feet below Camp 3. Broad, twenty-seven, was one of this expedition's lead cave divers. Normally Stone would have been happy to see him. But the expedition had not been going as well as he had hoped. Many days of grinding underground labor and what he considered to be halfhearted efforts by some of the crew, despite the example of his own near-frantic pace, had been exacerbated by his awareness of the coming rainy season and the fear that he would not be able to mount another expedition until who knew when. All of this had eroded his patience, so he was not happy about being yanked out of much-needed sleep. Nevertheless, his first thought was: *This can't be good.*

It was not. "Ian is missing. We need to mount a rescue right now," Broad announced.

IAN ROLLAND, A TWENTY-NINE-YEAR-OLD Royal Air Force sergeant, was Broad's diving partner. Rolland was diminutive—about five foot six and 145 pounds—but tremendously strong. A husband with three young children, he was good-natured and even-tempered, with the rare ability to disagree, even under great stress, without being disagreeable. He was no cream puff—Rolland's edged sarcasm could lacerate as well as elicit laughs—but as often as not he was the butt of his own gibes. All these traits were invaluable on demanding expeditions, where people who might or might not like one another were squeezed together, stressed and endangered, for weeks on end.

Rolland was a mechanical as well as psychological asset to the team. His job in the RAF was to maintain jet fighters, giving him an unusual feel for finicky, high-tech equipment like redundant, computerized experimental rebreathers. Ian Rolland had brought only one liability to this expedition: a year earlier, he had been diagnosed as an insulin-dependent diabetic. That would have kept most people from diving at all, let alone cave diving, but not this intrepid young Scot, who had learned to manage his diabetes with insulin and a proper diet.

A Miami native, Broad had more diving experience than any other expedition member, experience that included work as a stunt diver for filmmakers. He was slim and often had a growth of scraggly red beard sprouting from cheeks toughened by years of exposure to strong sun. Broad was also something of a superachiever. He was a U.S. Coast Guard–licensed captain, an EMT, a hyperbaric chamber operator, and was pursuing a Ph.D. in anthropology at Columbia University. For all his accomplishments, though, Broad was no stuffed shirt. Irreverent by nature, he loved to crack jokes and fire off smart-aleck retorts. Stone thought of him as the team wiseacre.

Rolland and Broad had not met before this expedition, but they became close friends quickly. The team's youngest participants, they shared a fondness for barbed humor and skill at using it to defuse stress. Each recognized the other as a master of this dangerous game, producing mutual respect. Given its hazards, confinement, and exhausting demands, caving was one of those activities—like combat, police work, and mountain climbing—that could turn people into raging enemies overnight. Or, as with Rolland and Broad, it could quickly forge friendships that otherwise might have evolved only over years.

The two of them, along with three other expedition members, had begun diving from Camp 5 on March 23. Their goal—the goal of this entire, massive expedition, in fact—was to find a way through San Agustín Sump, the flooded tunnel that had frustrated all attempts at penetration since 1979. This was supercave diving at its most challenging. For starters, living conditions at the divers' deep camp were horrendous. Here the cave's walls dropped straight into the sump's water. With no horizontal surfaces to occupy, the cavers suspended several red nylon, aluminum-framed platforms, called portaledges, from bolts driven into Huautla's walls. These rigid, hanging platforms, like those used by mountain climbers, were no bigger than the door to the average home. The divers lived on these spray-slickened shelves for days. Their bathroom was a plastic garbage bag. Some, like Broad, hung hammocks from the rock walls, but that was like trying to sleep in a body bag.

In the saturated atmosphere, the platforms were slippery and easy to fall from. Needing to relieve himself one night, Broad crept to the platform's edge and let fly into the stream below. To conserve carbide and batteries, he left his lights off and edged cautiously back to his hammock in the dark. Thinking he had arrived, he sat down, but his dead reckoning was off. The hammock spun and tossed him out. His head smashed into the cave's jagged wall. Stunned, flailing, he fell off the platform. In a move straight out of an Indiana Jones movie, he managed to grab one of the ropes on which the portaledge hung from the cave wall. Dangling there in the dark by one arm over the water ten feet below, Broad screamed for help, but the waterfall's roar drowned out his cries. With his grip loosening, Broad realized that he would have to save himself, and quickly, or fall and be swept away into the cave's black maw. He began swinging back and forth, waving his free hand around in the dark, and by sheer chance grabbed one of the other ropes from which the platform hung. With the last of his strength Broad dragged himself back up and flopped onto his belly, gasping and shaking, dizzy with pain, stunned by the fact that he had almost died in this supercave, not from diving but from falling out of bed.

Actually, sleeping was less of a problem than it might have been because no one really slept at Camp 5. Some 50 feet upstream a tremendous waterfall crashed and thundered, its roaring amplified by the rock chamber. It was like living next to a rocket engine with no off switch. To hear each other, Camp 5 occupants had to shout at the top of their lungs, which soon produced sore throats and laryngitis. The waterfall also kept Huautla's 64-degree air constantly saturated, making the camp a cold-shower room where the water never stopped running. Given all this, even brief stays here were exhausting and debilitating.

And all that was before the divers even got into the water. Once they did, the visibility was low, 5 to 8 feet, and at 64 degrees, the water was cold. The sump twisted and turned like a maddened worm—it was just hellish, one diver said. Underlying everything was the knowledge that if one of them suffered any of diving's myriad ills—the bends, oxygen toxicity, gas embolism, collapsed or ruptured lungs—or got hurt falling off a portaledge, help was so far away that they might as well have been on the far side of the moon. By the morning of March 26, these stressors had wrung out the other three divers, coolheaded veterans all. They headed for the surface, leaving Broad and Rolland alone.

At about four that afternoon, Broad helped Rolland slip into the sump's gray-green, chilly water. Diving with the rebreather, now called the MK-IV, was not a matter of hopping into the water and quickly submerging, as with traditional scuba gear. Because the rebreathers were so complex, and to avoid a repeat of an accident like Brad Pecel's, divers had to run through a long, detailed checklist similar to those pilots use to preflight their aircraft. A two-person job, it took more than fifteen minutes to complete.

Finally, Rolland rotated a valve on his black Delrin mouthpiece 90 degrees, activating the unit's gas-recycling system. With air flowing, he submerged and began the half-hour, southward swim to a chamber, called an airbell, that Broad had discovered the previous day. Broad himself settled down in Camp 5 to await his partner's return.

In the sump, Rolland followed white parachute cord that Broad had placed the day before during his dive. Swimming toward the airbell, intent on finding what lay beyond, he was surely having one of the great dives of his life. It was *quiet.* The rebreather emitted no bubbles, only the barely audible hiss of inspirations and exhalations. He was comfortably warm in his wet suit. The visibility wasn't great, about 8 feet max, but he had experienced worse.

At four-thirty, the rock ceiling above Rolland's head began to slope upward at a gentle angle. He followed it and soon saw the silvery reflection of the water's surface. He broke through and crawled on his hands and knees up onto a muddy sandbar. He was 1,410 feet from Camp 5. The air felt warm and muggy. His lights revealed that the sausage-shaped passage was 40 to 50 feet wide, 40 feet high, and 300 feet long—the length of a football field. Narrow sandbars ran like stripes of mustard down the chamber's length. Both water and air were still and, except for his breathing, the airbell was silent. To conserve precious gas while on the surface, Rolland disabled his rebreather's automatic oxygen injector. Doing so would save a few dozen breaths' worth at most. But down here a few dozen breaths could keep you on the right side of death's door, and for Rolland, a man made thrifty by heritage and meticulous by profession, it was second nature.

Rolland's MK-IV rebreather was the lightest version yet, but it still was no feather at ninety-five pounds. His remaining gear added another forty-five, giving him a total load almost equal to his body weight. It was an enormous burden, considering that the rule of thumb for fit backpackers is that you can handle one-third of your body weight. Trudging the length of the airbell while carrying his fins and walking through mud required immense effort. At the airbell's far end he rested, but not for long. His computer recorded a total between-dives interval in the airbell of just twelve minutes. At four forty-six he eased back into the water. For some reason, he did not reopen the rebreather's automatic oxygen feed. With traditional scuba gear, such an error would have been impossible. That much simpler equipment offered just two options: air or no air. Conventionally equipped divers assure themselves of the former by taking several test breaths before submerging. The more complicated rebreather, however, had more options, one of which could be, by mistake, breathing gas without the mix of oxygen.

Why would Ian Rolland, a world-class player of this deadly game, fail to make this essential reentry adjustment? It's impossible to know for certain. He was excited and fatigued, and may have been suffering exercise-induced hypoglycemia. For whatever reason—and it was probably a combination of all three—from now on, every breath would reduce the level of oxygen in his blood and increase its carbon dioxide level.

Back in the water, Rolland turned left and swam about 30 feet to the sump's east wall. Here the tunnel was not fully flooded, so he floated on his belly and tied off the guideline to a rocky projection just 2 feet below the surface. Then he turned right and swam along the wall for 50 feet, where he tied the line off again.

He pushed off and began to swim south once more, but then something happened. It hit hard and came on in seconds—that much is certain. It might have been hypoxia, a shortage of oxygen. But the most likely culprit, given his diabetes, was severe hypoglycemia, a sharp drop of his blood sugar, that could lead to insulin shock.

Many descriptions of hypoglycemia attacks exist, provided by victims who survived them. If—and it's a big if, for reasons that will soon become apparent—Rolland experienced that condition, this is what it might well have felt like: His muscles turned flaccid and weak and his vision blurry. He experienced flaring anxiety—rare for him on a dive. His heartbeat accelerated and his breathing rate soared as he turned back toward the sandbar and struggled into the dimming tunnel of his own light beams. He was only 50 feet from the airbell's sandy beach. He could make it that far. Surely he could make it that far.

SIXTEEN

BY 10:00 P.M. ROLLAND HAD NOT returned to Camp 5. That was the agreed-upon time when Broad would make the steep, hard climb back up to Camp 3 for help, which he had done. Now, in the early-morning hours of March 27, Stone and Broad gathered the others in Camp 3: Don Broussard, Rob Parker, Jim Brown. The group also included Stone's girlfriend, Barbara am Ende, thirty-four. She was tall and blond and slender, with a very attractive face that her practical Dutch boy haircut (less likely to get tangled in rappel racks and mechanical ascenders) emphasized rather than disguised.

Am Ende and Stone, forty at the time, had met during a West Virginia cave rescue in 1992. The attraction had not been instant. Stone's reputation as a stellar expeditionary caver was already well established, and when he learned that a woman would be part of the rescue team, he expressed his skepticism directly to am Ende. He didn't go so far as to kick her off the team, however, which turned out to be a good thing because am Ende and Stone were the first rescuers to find and help the injured caver.

That impressed Stone. The next morning, he lingered in camp to chat her up. Well, sort of. Bill Stone wasn't much of a chatter. His pickup line was really a series of short, sharp questions about her caving experience, delivered almost as though he were interviewing her for an expedition. (Which, in a way, he was.) For her part, am Ende was angered rather than attracted, but she could not help being impressed by Stone's energy, strength, and caving skill. A kind of electric charge seemed to surround Bill Stone, felt by all, men and women alike, who came in contact with him. It was, for lack of a better word, *exciting*.

Perhaps understanding that Lothario was not his best role, Stone secreted a business card in one of am Ende's packs. She found it after arriving home, and understood that she had at least passed Stone's first level of scrutiny.

There were others, of course. Stone was grieving the loss of his family—he and Pat were separated, their divorce still in the works—and did not want to go through that again. Any woman he became involved with now would have to keep up with him on the surface as well as in caves. That meant, first of all, being a competent caver. It also meant being fit—very fit. Stone knew that expeditionary caving, and especially deep-cave diving, demanded extreme fitness, and he worked hard to stay in shape, running and cycling and weightlifting. Am Ende passed that test, too. She was trim and athletic, riding a bicycle from her apartment to classes every day, and was a regular runner as well.

Before long they were both commuting. Stone drove down to the University of North Carolina at Chapel Hill, where am Ende was working toward her Ph.D. in marine geochemistry. Doing her part, am Ende made the five-hour trip north to Stone's house—not the former family residence—in Gaithersburg, Maryland, a suburb of Washington, D.C., where Stone still worked for the U.S. government. As the months passed, their relationship deepened. Barbara am Ende was passionate about caving, could just about keep pace with Stone, and was lovely, to boot. Am Ende found the tall, strong, utterly confident Stone fascinating and stood somewhat in awe of his caving exploits. As another plus, her cooler, more laid-back personality made a nice complement to his intensity.

In late January 1994, she pulled up stakes and moved from Chapel Hill to live with Stone. Less than three weeks later, they left together for Huautla, Stone to lead his most massive and ambitious expedition to date, am Ende to encounter her first supercave.

AS IT TURNED OUT, AM ENDE was the only woman on the Huautla team, an inclusion that caused problems from the outset. Her presence disturbed some of the men, especially Steve Porter, who was no less outspoken for being a rookie. During one team meeting, which did not include am Ende, Porter said, "Let's be honest. She wouldn't even be here if she weren't Bill's girlfriend."

But that was not entirely true, and am Ende had her defenders. One was a more experienced cave explorer named Tom Morris, a biologist from Florida.

"She's like some dream Amazon," he said. "Some tall, blond geologist caver who dives his rebreathers. Good for friggin' Bill."

Am Ende was a cave diver, albeit a newly minted one, having earned her basic scuba certification just a year earlier. She was also the first in a series of younger, very attractive women Bill Stone would take with him to the great caves. The habit invited another comparison between Stone and Reinhold Messner, who brought beautiful young women along on some of his memorable climbs. During his greatest of all, the solo, oxygenless ascent of Everest in August 1980, his lover Nena Holguin held silent vigil for him in their tent and was there to help when he returned, utterly spent.

BLEARY AND BEDRAGGLED, THE CREW AT Camp 3 listened to Broad's explanation. The seven stood around, wet, muddy, dulled by fatigue, prison-pale, smelling of mold and excrement, the carbide lamps on their fiberglass helmets burning white circles in the darkness. It was one of those situations for which no amount of years of experience or training could fully prepare them. Down here, as in life, the line between happiness and horror was thin indeed. But horror up top can be mitigated: by family, friends, clergy, police, doctors. Down here, *they* were the mitigation. Getting help would not be an option.

That was true for two reasons. They were at least a day's climb from the surface. News of Rolland's absence had reached them no faster than Broad could, and a request for help would make it to the top no sooner than someone could climb up there and deliver it. Mountaineers had been using radios since the 1960s, and divers had had "voice-comm" capability well before that. But radios are useless for surface-to-depth cave communications because their waves cannot penetrate solid rock. As late as 2002, information in caves still traveled no faster than in the ancient times before humans rode horses—

walking speed, in other words. And not even *running* speed, because caves don't lend themselves to a faster gait. Stone once remarked that from deep camps, "You would send out this laundry list. Guys would carry it back to the next camp. 'Did they say *this* or *that*?' So by the time the answer came back in, you didn't always get what was on that list. And you just lost a week."

THE CRISIS PRESENTED STONE WITH a Hobson's choice: further endanger Rolland by delaying a rescue attempt or risk others' lives by rushing one. Talking to the team, using his engineer's logic to analyze options, he explained. If Ian was dead, game over, no need to rush. Cold, but true. If he was alive but had not returned, he obviously needed help. But he could be alive—undrowned, in other words—only if he was out of the water, stranded but safe for the time being on some dry land. If that were the case, they must rescue him. But to do that, one or more of them would have to dive the quarter-mile sump, and to do *that* they would have to assemble the expedition's other rebreather, a multihour job devilishly challenging even above ground, with good light and clean surroundings. Down here, screwing up just one tiny task out of hundreds could mean death for the next user. Assembling the apparatus was no job to rush, especially not by people stupefied by exhaustion. They needed several hours of rest first.

Appalled, Broad objected—strenuously. He and Ian had become close, and, familiar with diving's hazards, he could visualize his friend alone in the cold darkness, suffering the agonies of decompression sickness ("the bends"), or lying immobilized by some surface injury, or sickened and dying from his diabetes. Broad simply could not accept delaying a rescue attempt, risk be damned, and said so. But it was not his decision to make.

Stone determined that they would all sleep for a few hours, then go down to mount a rescue from Camp 5. As he had said, if Ian had already drowned, nothing could be done, though that possibility was not a nice thing to contemplate. Drowning is a cruel way to go. It throws two of the body's most potent self-preservation reflexes into competition. Trapped underwater, you hold your breath as long as possible, with the urge to breathe growing from a whisper in your chest to a scream in your brain. As the carbon dioxide in your bloodstream builds up, you start to jerk and spasm. Gray fog closes down your peripheral vision. Ultimately, it is all a matter of chemistry. With your vision down to points of light, your fists clenched and toes curled as if in orgasm, your mouth opens not to scream but to inhale involuntarily. Most drowning victims reach this point in ninety seconds, give or take, which is very long indeed in extremis. Finally, your lungs fill and you become negatively buoyant, floating slowly down, staring at eternity. There may be no good ways to die, but some are worse than others.

AT FIVE THE NEXT MORNING, March 28, barely rested, Bill Stone, Kenny Broad, and three others climbed down to Camp 5, where they reassembled the expedition's second rebreather. Shortly after noon, Broad geared up and disappeared into the sump. He reached the airbell at 12:45 P.M. He explored the passage with his light and found footprints leading onto the sandbar. He followed them to where the sandbar sloped into clear water about 10 feet deep. Broad entered the water and swam, looking down at the bottom through his mask, for about 50 feet on the surface. Then his circle of light surrounded Ian Rolland, lying still on the sandy bottom. His mask and other equipment were in place, undisturbed. A red light was blinking on his rebreather's buddy console. There was no evidence of struggle.

Because his was a reconnaissance rather than an investigation dive, Broad had no underwater slate to make detailed notes on, so he left everything untouched and headed back to Camp 5. Bill Stone, awaiting his return, spotted his light approaching. Broad surfaced, spat out his mouthpiece, and took off his mask.

"Ian drowned," he said.

Speaking first and foremost of himself, Stone had once remarked that if some people die making faster pushes to the moon and Mars that are necessary for successful space exploration, well, no big deal—there are lots more where they came from. Those who know Stone, whether they like him or not, will tell you that he has the strongest will they have ever encountered. Such Shackleton-class resolve stood him and all the others in good stead just now. Compartmentalizing grief, he set about organizing a recovery. Keeping Kenny Broad with him, he sent the other three up for help. It was going to take a large team to get Ian Rolland out of the cave, but he *would* come out. Stone well remembered the Chris Yeager controversy. Before Ian could be carried out of the cave, he had to be brought back through the sump. Stone knew it would be the most difficult and dangerous thing any of them had ever done. And, since he was both the expedition leader and one of its most experienced divers, the job would fall to him.

SEVENTEEN

CAVE DIVING, BY ITSELF, IS EXTRAVAGANTLY dangerous, but recovering dead bodies from caves is even worse. For one thing, divers usually die not in benign places but in the more dangerous parts of caves. In addition, thrashing around in his death throes, a diver often becomes entangled in his own safety lines, which, in addition to the other guidelines strung in caves, create a deadly web waiting to snare rescuers. A live diver can squeeze and wriggle through perilously tight passages, some of which require doffing a tank, pushing it through, following, and then donning it again. Working a dead body back through such places (the need for forensic investigation makes it crucial to retrieve gear as well as body) is hideously difficult, exhausting, and invites damage to the rescuer's own gear. Such contortions inevitably stir up huge amounts of silt, turning almost every recovery into a zero-visibility encounter, which increases all the other risks by an order of magnitude.

At about eight-thirty the next morning, Stone donned his usual diving gear, plus extras for this specialized mission: multiple carabiners to attach Ian's body to his own, and straps to bind the dead man's arms and legs if rigor mortis had stiffened them awkwardly. He dove through the sump, crossed the sandbar, and found his friend's body. Rolland was lying on his right side, facing toward the passage, about 50 feet into the second sump. His red plastic line reel was lying on the bottom, 8 feet from his body. As Kenny Broad had noted, the death scene did not give an impression of drowning, which is often accompanied by evidence of frenzy—mask torn off, gear in disarray, hands lacerated, the diver entangled in his own safety line, disturbed silt resettled on the body. Stone saw none of that here, either. Rolland's mask was in place. His hands were unscathed. The rebreather's mouthpiece was hanging loose, emitting a slow, peaceful stream of bubbles. Both backup regulators were still properly secured and functional. Four of the five tanks he carried were full or nearly full.

So what *had* killed Rolland? Hypercapnia, an excess of carbon dioxide in the system, was one possibility, but unlikely. For that dive, Rolland had meticulously repacked the rebreather's scrubbing canister, which removed carbon dioxide from the recycled gas. Insulin shock induced by hypoglycemia was another possible culprit. Crossing the sandy airbell floor after a stressful swim through the sump while loaded down with 140 pounds of equipment would have been a likely way to induce insulin shock. Just the anaerobic slog by itself could have had dire consequences.

Finally, there was hypoxia, a shortage of oxygen. Ultimately, hypoxia causes death by halting all normal metabolic functions, but the diver's brain cells are more at risk than any others in the body. Because the brain is the organ first and most affected, a diver can become unconscious before recognizing any other symptoms. As in the case of hypercapnia, hypoxia can kill before the victim even understands that anything is wrong. Data downloaded later from the rebreather's computer strongly suggested that hypoxia had not been at fault. But that information was not available to expedition members at the time.

Stone recorded information on waterproof slates, then hauled Rolland and his gear and himself, a total load of over six hundred pounds, up onto the sandbar and back to its far end. To retrieve the body, he would have to clip it to his chest harness, putting his face inches from Rolland's for the duration of the dive, his bright helmet lights illuminating everything. Noel Sloan had recovered many bodies from caves. He'd once told Stone, *When you have to do it, turn the wet-suit hood around.* Stone took Sloan's advice.

With the body fastened tightly to him, Stone crawled into the water and set off. Managing buoyancy in open water is scuba's toughest discipline to master. Managing the buoyancy of two bodies, one of them dead, in a cave sump is nearly impossible. Stone yo-yoed erratically between the sump's floor and ceiling, stirring up more silt all the time. Twice he lost the guideline. In a flooded, silted-out tunnel 90 feet wide and a quarter-mile long, this easily could have been a death sentence. Each time, Stone managed to relocate the little string. Sometimes fortune really does favor the bold and the brave.

At 11:30 A.M., he resurfaced in Camp 5. Immediate examination of the rebreather, as well as a later, much more thorough laboratory exam, all but ruled out MK-IV malfunction as the cause of death.

Ian Rolland's diabetes was a more plausible explanation. He had been gone for a long time and might have already been mildly hypoglycemic when he entered Sump 2. But if that *was* the culprit, it killed him with astonishing speed, so quickly that he did not even have time to eat a Power Bar—the two he carried were still in his wet-suit thigh pocket when Stone found him. Frustratingly, not even that conclusion could be drawn with certainty because the body's advanced decomposition prevented testing for insulin shock. The medical examiner's verdict only added more shadows to a death already shrouded in mystery: the autopsy revealed *no water* in Rolland's lungs. The official cause of death was listed as "asphyxia due to immersion in water," a verdict no more satisfying than the one that had been rendered following Rolf Adams's accident.

It took five days to haul Rolland's body from the cave, wrapped in a plastic tarp, dripping fluids of decomposition onto its handlers, its Gorgon head shrouded in a wet-suit hood. Word of the death quickly circulated through the expedition and then the region. And well before the recovery team surfaced, people were speculating that the experimental, complex rebreather had killed Ian Rolland, one of the world's best cave divers and one of the expedition's most popular members.

The death was bad news for an already troubled team. For weeks, Stone had been driving himself and everyone else relentlessly. Depending on who was commenting, he was either "hyperinsensitive" or admirably committed to the mission, and more people were inclined to the former. Team members had taken to calling Stone "the Bulldozer," as well as less flattering things, and bridled at his attempts to make them match his own frenetic pace. There was also Barbara am Ende's presence, which for some was like the grain of sand in an oyster, a constant source of irritation that could not be reduced.

For his part, Stone believed that people who had benefited from his core team's immense preparatory work had been slacking for quite a while, and that angered him. Regardless, Rolland's death was the final blow. People, including some whose agreements called for them to stay until the expedition's end, began talking about pulling out because of the death. To Stone, using a comrade's death as an excuse to quit was reprehensible. But it didn't feel that way at all to a number of the team members. With defections threatening, it was beginning to look like a repeat of the 1984 Peña Colorada mutiny.

EIGHTEEN

THINGS BEGAN ESCALATING FROM DESERTIONS to attempts to derail the whole effort. Rob Parker, a respected British cave diver and a friend to both Ian Rolland and Stone himself, was terribly disturbed by Rolland's death. He was, in the estimation of Stone and others, a true superstar, and he went around camp repeating to anyone who would listen, "Ian himself said, 'Somebody's going to die on this machine.' " Parker wanted the expedition suspended right then. Stone could understand Parker's grief. But he thought that the idea of ending the expedition to honor Ian Rolland was "horseshit" and refused to do it.

His decision to continue was not well received by a team stressed to the breaking point and being sown with seeds of discontent by Rob Parker—and others. Parker was not the only one saying openly that the rebreather had killed Rolland. Bill Farr (co-discoverer, with Carol Vesely, of Cheve) was making similar accusations. Farr had been with the expedition at the beginning but had removed himself early on from the dive team. He had spent most of the expedition over at the Cheve resurgence, and had returned to Huautla only in time to witness the emotional procession of cavers and locals bringing Rolland's body up to the village from the cave entrance. Some members of the expedition, including Barbara am Ende, felt that Farr had contributed little to the overall effort, and certainly not enough to justify speaking as though from some bully pulpit.

But Farr, who gave away eight inches and fifty pounds to Stone and was even smaller than am Ende, was nevertheless himself a stubborn alpha. A veteran of many expeditions, he was appalled at the deterioration—exhaustion, anger, eroded morale, and the death of a member—he perceived in this one. He felt that continuing would probably add catastrophe to tragedy. Things finally came to a head one morning in the base camp cookhouse, a one-room, vacant village home the expedition had rented. It had bare wooden walls adorned with fading pictures of Jesus, a concrete floor, and a corrugated metal roof so low Stone banged his head if he stood up straight without thinking. Two benches and a long plywood table that Ian Rolland had built occupied the center of the cookhouse. It had become the expedition's de facto headquarters, a place for cooking and eating but also for parties, bull sessions, and important meetings.

On Saturday, April 2, an exhausted Stone slept late. He planned to hold a meeting of the entire team the next day to discuss the future of the expedition. Am Ende left Stone asleep and walked over to the cookhouse. Entering, she found another meeting already in progress. Bill Farr was declaiming to a group around the table, not suggesting but *telling* them that the expedition was over.

Furious, am Ende lashed out at the Californian: "Excuse me, but what in hell are you talking about? There is no evidence that problems with the rig caused Ian's death. There's no evidence for that."

"Well, but we've had problems since Florida," Farr countered, in what am Ende heard as a condescending tone. "And Noel told me that he almost went hypoxic on his last dive."

"Who's *we*, Kemo Sabe?" am Ende snapped. "I don't recall seeing your smiling face at Ginnie Springs [a training site]. The rigs were working fine. But you don't know that, because you weren't there, were you? As for Noel, he didn't run out of oxygen. He had hypercarbia because he forgot to change the damn lithium hydroxide canister because Noel . . . well, because he's Noel. That was operator error, not a problem with the rig."

Barely pausing for breath, she rushed on: "And Noel completed that dive without incident, didn't he? That's because all he had to do was switch to the offboard bailout system. If that had somehow failed—which it didn't—he could have grabbed one of the bailout bottles stashed in the sump. So there were lots of bailouts. Ian knew all about the bailouts. But you wouldn't know about them, would you? *Because you didn't dive, did you?*"

Everyone sat in stunned silence. Even am Ende was a bit surprised by her outburst. But she wasn't finished. "The fact is," she continued, "we don't know what happened. We'll review the data that Bill and Kenny collected and we'll adjust our procedures, if necessary, so that every dive will be a safe dive."

"There aren't going to be any more dives. This expedition is over," Farr decreed. Am Ende was astonished at what she considered the man's crude and ill-timed grab for power.

"You don't know that," am Ende shot back. "That's not your decision to make. We'll have a team meeting tomorrow or the next day. And the team—that means the people who actually do the work around here—the team will decide whether or not to continue with the expedition."

But Farr would not be deterred. "They terminated the Cheve expedition after Chris Yeager died," he retorted. "This project needs to be stopped, too."

Am Ende could only stare. *Who the hell does he think he is?* she wondered. Disgusted, she spun on her heel and stalked out of the cookhouse. Not long after, a young British caver named Mark Madden came out to join her.

"Quite an immediately dislikable fellow, that one," he chuckled, smiling sympathetically.

Am Ende agreed wholeheartedly, and Madden's quip helped dispel some of her lingering ire. She laughed, and Madden went on his way. Before long, though, another cookhouse witness, the diver Rob Parker, walked up.

"I'm on my way into town to talk to the local authorities," he told her.

"*Please* do not tell them that the expedition is over," am Ende beseeched him.

"In reality, it might be," Parker said.

So it's not just Farr, she thought. "It's not Bill Farr's or Rob Parker's or my decision as to the fate of the expedition. That's a decision the whole team has to make," am Ende declared.

Parker clearly disagreed, and went on his way without another word.

THE INCIDENT LEFT TEMPERS SIMMERING on both sides. Stone convened the meeting as planned the next day, April 3, which happened to be Easter

Sunday. He tried to hold the team together with a speech. His personal philosophy, that they were engaged there in something far more important than "adventure," was at its core.

"Three or four hundred years ago . . . ships would often lose thirty percent of their crew in the course of a voyage," he reminded the cavers. "The difference between us and them is that our society now places so much importance on life."

The message fell on mostly deaf ears. For many of the team members, deep caving *was* an adventure, with science second at best. They just could not see things as Stone did. Nor could some local residents. A few days later, Stone met with the nearest village's three town officials to report on the status of the project. Two of the men were well-educated teachers and businessmen. The third was an old, traditional Mazatec. After hearing Stone out, he said, "I will tell you why this good man died. You did not seek permission from Chi Con Gui-Jao. You have been arrogant. And for this you must pay the price." Chi Con Gui-Jao was the spirit that lived in Huautla. The Mazatecs knew this to be true, and believed it as fervently as Christians believed in the Resurrection.

Many, perhaps most, members of the expedition agreed. Overworked, exhausted, devastated by tragedy, and leery of the rebreathers, members began packing up and leaving. As April got under way, there were five team members left: Bill Stone, Barbara am Ende, Noel Sloan, Jim Brown, and Steve Porter. Stone blamed himself for the team's dissolution, as he had for the 1984 mutiny. If he had learned anything, it was that the guy who's in charge has to keep the focus. And that meant inspiring the troops with action. If *I* do it, they will follow, he believed. Or, perhaps more accurately: If I do it, they *should* follow. That was how leaders should lead, and that was how he led.

On an earlier expedition in nearby Puente Cave, Stone had had a revealing exchange with a younger male caver. Thinking to inspire the younger man, Stone told him that the descent they were about to undertake—together—would be the toughest, most brutal trip he would ever do. *Ever.*

"This will make a real caver out of you," Stone said, his tongue at least partially in cheek. Thinking that he had whipped the fellow into a frenzy of anticipation, Stone was surprised when the neophyte asked, "But is it going to be fun?"

Fun? *Fun?* "Of course not," Stone said. By 1994 he had long since stopped thinking of this stuff as fun. No, it would be goddamned bloody awful, he told the young man. This wasn't a vacation. It was exploration, on the last great terrestrial frontier, in the name of science.

"Well, why would I want to do *that*?" The young caver seemed as genuinely mystified as if Stone were speaking a different language. Which, in a sense, he was. Bill Stone was just as confused. He had been doing his share and more, and would continue to. What was wrong in that other guy's head?

The exchange revealed something valuable. Stone would have made a superb general in the Civil War, where leaders led from the front and took the first bullets. For them it was a matter of faith that coming close on their heels, without pause or complaint, was how their followers would follow.

Scientists have long known the fascinating fact that the human eye has a blind spot in its field of vision, right on the point of the retina where the optic nerve leads back into the brain. Even more fascinating is psychology's awareness that we have a kind of emotional blind spot as well. Awareness does not equal understanding; there is debate about why we have emotional blind spots, but none about their existence. No blindness is benign. Reduced to essentials, emotional blind spots cloud judgment, flaw decisions, and damage relationships.

Stone led by challenging people because that was what best motivated *him*. For the sake of a mission, he himself was willing to challenge anything, up to and including universal laws. The quote that opens Part One of this book explains Stone's philosophy: If a universal law gets in the way, hell, bend the bastard. Challenge was one way, maybe one *great* way, to inspire, but it was not the only way. If Bill Stone had a blind spot, it was failing to understand that every lock requires a different key.

Two differences distinguish most people from the Bill Stones of this world. One is that a big majority are happy living and working quietly in the shadows. Bill Stone and the few like him spend much of their lives in the spotlight. The bright glare invites applause, but it exposes flaws with equal intensity.

The second difference is that our actions rarely invite injury and death, while theirs not infrequently lead to both.

NINETEEN

DEATH WAS VERY MUCH ON THE minds of Jim Brown and Steve Porter just then. Neither would dive again on the rebreathers. Sloan might, but his commitment was tenuous at best. Only one person was both able and willing to do that with Stone. "Able" is but half the equation, of course, "willing" being the other half. Barbara am Ende, though a seasoned dry caver and an experienced open-water diver, had logged only about thirty cave dives, and those in much less lethal environments than Huautla. Open-water divers are not considered truly experienced until they've logged several *hundred* dives, and that kind of diving is to cave diving as ballet dancing is to bullfighting. What's more, because they only had two rebreathers at the sump, and because Bill's had been configured for his use, Barbara had to dive with Ian's "death rig," an option already rejected out of hand by three very good, very brave cave divers.

If Barbara am Ende had loved only caves or only Bill Stone, this whole thing might have gotten to "can" and stopped short of "will." But she loved both. And so this remarkable woman, whose last name in German means "to the end," agreed to keep diving Huautla's deadly sumps with Bill Stone, beyond all hope of rescue, on the suspect rebreather Ian Rolland had been wearing when he died.

THE ODDS STACKED AGAINST AM ENDE and Stone were daunting. For one thing, they were depleted, like fighters who had just gone twelve rounds but were being forced to fight again without leaving the ring. And though they had maintained their mission focus, working hard to remain stoic for the benefit of the others, Ian Rolland's death had shaken both of them. So had the attacks on Stone by members of his own expedition, some to his face, others behind his back. These disturbed am Ende as much as Stone—perhaps more.

Then there was the matter of diving. Barbara am Ende had practiced with the experimental, massively complicated rebreathers in the springs of Florida, but those springs were bathtubs compared to Huautla. The rebreathers had taken punishing use for weeks ("We beat the shit out of them" was how another expedition diver put it), and one might have just killed a world-class cave diver. That still very open possibility was strong enough to have effectively ended the expedition. The other side of the San Agustín Sump was terra incognita.

Finally, rescue was not a realistic option if catastrophe struck—which, given what am Ende and Stone were about to do, seemed more than probable. In addition to all the routine minor injuries and illnesses involved in supercaving, cave diving invited an array of ailments as unpleasant as their names implied: decompression sickness, arterial gas embolism, nitrogen narcosis, shallow-water blackout, pneumothorax, oxygen toxicity, seizures, and more.

If either or both explorers were incapacitated, even by something as minor as a badly sprained ankle, any potential rescue team would first have to reach Camp 5, then dive the San Agustín Sump, Sump 2, and any others the pair might have moved through. They would have to locate one or both victims, bring them back through the sumps on scuba gear, and then transport them 6 miles and 4,100 vertical feet to the surface, negotiating more than ninety rope climbs on the way, some hundreds of feet high.

The simple fact was that that rescue would be virtually impossible from where the two were going in Huautla. In that regard, their excursion would be unique. There was no other place on earth so remote. Helicopters make extractions from jungles, deserts, oceans, and even the highest mountains possible. Similarly, submersibles enable rescues from deep beneath the sea. No such technologies could come to the aid of these supercave explorers. One of their friends, the great American caver Mike Frazier, said of injuries deep in a supercave, "If you get hurt bad down there, chances are good you're not coming out."

True enough. Thus they were embarking on the caving equivalent of an extreme free solo first ascent in rock climbing, something on the order of El Capitan with no protection, rope, or belays.

THEY WANTED TO START IMMEDIATELY, but certain things had to be done first. Ian Rolland's body had finally been brought out of the cave on March 29, with two recently arrived Mexican policemen on hand to observe. On April 1, the police inspectors were still there when a touching memorial service was conducted in the village church. Rolland's body, well into decomposition, was then removed to the team's gear-storage room, where it would stay until arrangements could be made for its return to Scotland. Right after the memorial service, though, the policemen made an unexpected demand:

We must view the body! Take us to it now.

Noel Sloan, Stone, and others escorted them to the gear-storage building, where Rolland's body, still in its wet suit and wrapped in two plastic tarps, rested on a plywood table. Inside, astonishingly, one of the policemen produced a small toolbox and unpacked syringes, cotton swabs, red rubber gloves, and an X-Acto knife. These he handed to Noel Sloan.

Now you will perform the autopsy, he said.

A seasoned emergency room physician, Sloan had no aversion to gore and putrefaction, but he knew that a butcher-shop operation like this would reveal nothing of value about how Rolland had died. It might even invalidate Rolland's life insurance policy, leaving his family with nothing.

We cannot do an autopsy here, Sloan explained, speaking through an interpreter.

The two policemen huddled briefly, then replied: Very well. But we must see the body. It is required. When Ian's corpse had been carried out of the cave, it had been wrapped in the orange tarps, so the policemen had seen evidence of the body, but not the body itself.

You must understand, Sloan said. He has been dead and decaying for a week in the heat, sealed in this plastic cocoon.

The policemen stared at him, understanding but unyielding.

You really don't want to do this, Sloan said.

But they could not be swayed. Please proceed, one ordered.

Sloan sliced through the tarps one by one, then the wet suit. When he peeled back the tarps, vile fluids flowed over the table and onto the floor, splashing the Mexican cops with the juices of death. It was more than they had bargained for. They jumped back, gave the body a hasty once-over, and fled.

Then arrangements had to be made for returning the body to Scotland, and right after that Stone and am Ende had to spend a week helping prepare the cave camps for a photo team from *National Geographic*, a key expedition sponsor. Finally, on April 8 the two of them were able to slip down to Camp 5, allowing Stone to do a solo reconnaissance dive through the two sumps and briefly beyond. But immediately thereafter, they had to come all the way out of the cave again so that Stone could allay local unrest, fueled by Rolland's death, with a public slide show in Huautla village. More work for the photo shoot consumed another two weeks, until the *National Geographic* team finally headed back to the States on April 23.

Stone and am Ende were still determined to press on at all costs. Their stalwart friend Noel Sloan, unnerved by all that had happened, would keep going, too, but he was less enthusiastic than ever about diving. Steve Porter, still hanging around camp, was fuming and would not dive again. Reclusive Jim Brown was in a funk, living hermitlike in his van; no one was sure *what* he would do.

Am Ende wanted to dive beyond the San Agustín Sump, but she believed that Noel Sloan should go if he wanted to. He was older, had far more diving experience, and had worked with the ongoing Huautla exploration much longer than she had. But Sloan looked gaunt and exhausted. She could tell that he had been badly rattled by Rolland's death, and she suspected that, deep down, he didn't want to dive.

Noel Sloan was an unusual man, even by supercaving standards. Despite his courage, skill, and stellar cave work, there was a mercurial quality to him that caused Stone to think of him as an "emotional amplifier." When things went well, Sloan was ebullient, almost giddy. When they went badly, his morale plunged—even lower, sometimes, than the situation warranted.

There was more. For all his scientific and medical training, Sloan was superstitious. Perhaps he really was one of those people who feel things the rest of us cannot. He had experienced, after all, that shocking premonition of Rolf Adams's death at Jackson Blue Springs back in 1992. He placed great stock in signs and premonitions and refused to scoff at Native American beliefs about cave gods and evil spirits. Sloan did not just pay lip service to such beliefs—he acted on them. A "bad feeling" after his initial recon dives into San Agustín Sump had led to his early departure from Camp 5. Later, shaken by Ian Rolland's death, he secretly visited a local *curandero*, as the Mazatecs called their shamans.

The *curandero* would meet Sloan only on a night when the moon was full. Sloan found him in a nearby village, in a dim, dirt-floored hut. The shriveled, white-haired old man burned a fragrant tree-resin incense called copal, read the future in magic corn kernels, and chanted. Sloan joined in prayers for the cave gods' forgiveness and the expedition's safety. The *curandero* gave Sloan two sprigs of a sacred plant called the herb of San Pedro. Plant one of these at the cave entrance, he said, and carry the other one into the cave with you. In addition, he ordered, every person who goes into the cave now must carry garlic. Finally, he told Sloan that eating sand inside the cave would give him courage. Later, Sloan did plant the herb in a spot of sunlight by Huautla Cave's entrance, and he made sure that all five team members had garlic cloves in their packs before they descended on April 26.

They spent the next three days stocking Camp 5, and on April 28, the whole team, such as it was, came together at Camp 3. Soon, Sloan pulled Stone aside for an unnerving conversation.

"Before we came down in here," Sloan said, "I called my parents, and my in-laws, and my wife. I said good-bye to all of them." Given what Stone knew of Noel Sloan's premonition before Rolf Adams's death, it was about the most unsettling thing he could have heard.

It sounded to Stone as if Sloan had accepted the fact that he was going to die and had been putting his affairs in order. *This is not the Noel Sloan I know*, he thought.

"Look, Noel, you don't have to make a decision just yet," he said. "Why don't you sleep on it, and let's talk in the morning."

Sloan agreed, but his parting words did little to reassure Stone: "There are a lot of weird vibes going on right now."

The next morning, am Ende, Sloan, and Stone prepared for their last trip down to Camp 5. Before they headed out, Sloan spoke with Stone.

"I can't do it. I've lost my edge," Sloan said, appearing almost as unhinged by his own actions as by Rolland's death and the omnipresent danger. "You've seen the crazy things I've been doing. I'm not ready to make the dive."

A bit later, he spoke to am Ende. "You and Bill have been working as a team on this whole expedition," he said. "I think you should go."

Sloan's decision solved one problem but created another. The three of them knew that Porter and Brown did not think that am Ende should dive. Porter had stated unequivocally that he would quit on the spot if he learned that she was going to, and Brown would almost certainly follow him. But their help was desperately needed to finish stocking Camp 5. There seemed no other choice, so Stone, am Ende, and Sloan hatched a plot. They pretended that Sloan was going to do the dive. The five of them would finish hauling loads down to Camp 5, after which Porter and Brown would return to commodious Camp 3, leaving the conspirators alone.

With that agreed upon, Stone wanted to have one last conversation alone with am Ende, and he knew just the place. Halfway between Camps 3 and 5, they huddled in an alcove in the wall that, like a phone booth, afforded some privacy, if not security. This deep, Huautla was a violent place. Hard wind stripped spray from flowing water and spun it over black walls laced with bright white stripes of calcite.

"Are you absolutely sure you want to do this dive?" Stone asked.

"I feel like I've trained for this all my life," she answered. "I know I can do this. The line is there. I can handle the buoyancy." She had been in lots of big caves before. Like a veteran marathoner who's experienced hitting the twenty-mile wall, she recognized the physical and mental walls presented by caves in general, plus those this cave was now throwing at her. She was cold, wet, tired, disturbed by Rolland's death, and anxious about diving a strange sump. But she had experienced difficult things before, kept her cool, retained her mission focus, and kept going. Once she got moving, am Ende felt sure, her brain and body would warm to the task. But she was not going to make the dive alone. An extended recon beyond the sump would require a tremendous amount of gear and supplies, which Stone would have to carry. She turned the tables on him with a question of her own:

"Can you handle the gear?"

Stone had been thinking about that himself. "Yeah," he answered. "It's going to be a mother of a pack. As long as I can get it neutral, I'm pretty certain I can do it." An object that is neutrally buoyant neither rises nor sinks.

They had made their peace and would go forward. Stone was relieved, but also aware that he was shouldering a huge responsibility. He would not only have to manage all the challenges of his own diving, he would have to be hypervigilant about am Ende's as well, constantly thinking two or three moves ahead, monitoring her performance, anticipating problems, planning escape contingencies. He was comforted by the fact that am Ende possessed that one quality he valued perhaps more than any other: grace under pressure, the ability to ignore distractions, cut through confusion, and *focus*.

For her part, am Ende felt confident that she could deal with the physical challenges involved. More difficult, she knew, would be managing her fear. But she had not caved and dived this long without becoming expert at that. She understood going in that supercave diving was a dance with death; you either accepted that or you didn't play. Ian's death was unsettling, but rationally she knew that he had not been killed by any anomaly in the sump. It was a deadly sump, as most were, but not inordinately so. The water was cold but not debilitating, its flow was not brutal, its visibility was poor but acceptable, and the sump itself was not all that long or deep. She knew that Stone had made it over and back twice—once hauling a dead body, to boot.

That evening, Brown and Porter ascended to Camp 3; two days later, they climbed out of the cave for good, leaving am Ende, Sloan, and Stone by themselves in Camp 5. All seemed ready for am Ende and Stone to make their big push.

TWENTY

THE NEXT TWO DAYS BROUGHT a series of terrifying equipment failures that would have discouraged virtually anyone on earth—but not Bill Stone and Barbara am Ende. On the morning of April 30, he was preparing to dive when he heard a noise like a blown-up paper bag being popped. One of his rebreather's regulator diaphragms had ruptured and was gushing air. It could not be repaired.

Stone shot up a thousand feet, retrieved a replacement regulator, and zipped back down to Camp 5 in a matter of hours. Sloan replaced the blown unit with the new one. Later that afternoon, am Ende was in the water checking her own rebreather, the one that Ian Rolland had been using on his fatal dive. As part of the postmortem investigation, Stone had removed that rebreather's computer, which contained all the data from Rolland's last dive, and had replaced it with another. But the new unit did not calibrate depth correctly, making it useless.

Am Ende clambered out of the water and Stone went to work. Unfortunately, he made a wrong connection to the computer, using a high-pressure rather than a low-pressure hose, which completely blew out the unit's crucial depth sensor. This was too much for the increasingly unnerved Noel Sloan.

We need to end this thing now, he urged.

Stone and am Ende dismissed that idea out of hand. A search of Camp 5 turned up an extra sensor in, of all places, Ian Rolland's stashed gear. Stone installed the spare, but when he was finished it was late and everyone was exhausted. They decided to spend another miserable night at Camp 5. It turned out to be much worse than miserable, more like a waking nightmare. Am Ende's sleeping bag had gotten soaked during the day, and when she crawled in, it sucked away her body heat like a wet sweater in a strong, cold wind. Stone gave her his dry bag, then sat up a long time trying to dry hers over the tiny flame of their butane camp stove. Eventually he crawled into the still soggy bag, but without the distraction of any tasks, he became acutely aware of the waterfall roaring like a line of diesel locomotives at full throttle less than 50 feet away. The noise was beginning to erode even his superhuman resistance. Earplugs did nothing to help. In desperation, taking a tip from Kenny Broad, Stone rolled up his soft balaclava hat and pulled it down over his ears. Not much better. Wide-eyed and wide awake, he lay in his hammock and waited for sleep that would not come.

This place should be called Camp Fear, he thought.

THE NEXT MORNING, BOTH AM ENDE and Stone awoke groggy, cold, and exhausted. Running at full blast on its inexhaustible fuel supply, the waterfall was still roaring. Moisture collected everywhere, including on the soggy toilet paper they used during trips to the plastic-bag latrine. Those trips, and every other move on their slick platforms, required extreme care, lest they slip off, as Kenny Broad had done earlier. The pounding waterfall filled the air with perpetual mist that made their headlamp beams fuzzy, like car headlights shining into fog. It lent an eerie, nightmarish quality to the place, which their inescapable thoughts of Ian Rolland's death did nothing to lessen.

Stone and am Ende sat next to each other, not saying much; Stone could sense that something was amiss.

"How do you feel?" he asked.

"I'm not as enthusiastic as yesterday about doing the dive," she admitted.

Stone had felt that all was not right, but her statement still stunned him, coming when and where it did. This was, after all, the kind of moment that might present itself, for those lucky few willing to take ultimate risks, once in a lifetime, and for the vast majority, toiling slavishly out of the limelight, never at all. It was impossible to calculate all the hours and days and months invested to produce this one single opportunity, but Stone could feel their weight like lead on his shoulders, all the endless meetings on bended knee in corporate suites, the myriad hassles and obstacles to launching every expedition, the cajoling and massaging of local authorities, the disputes and desertions by those less committed, the abandonment by a disillusioned wife and disconsolate children, and the deaths of close friends. All of those and more he had endured and surmounted for this one, this irreplaceable chance to do something that forever after would force people speaking of Bill Stone to precede that name with the encomium "the great explorer."

There was something else, though, something even more immediate and troublesome: survival. He thought, *I do not want to do this dive with someone who might take me down with her.*

Stone knew well—better than any living practitioner, perhaps—what a lethal game of chance cave diving could be. Sure, you could train exhaustively and equip yourself with the best and highest technology available and follow established procedures religiously—and you could still die horribly. It had almost happened to Stone himself in 1979 in Huautla. It *had* happened to Rolf Adams, and Ian Rolland. Worst of all, just a month earlier, it had happened to the greatest cave diver of all time, Sheck Exley, who had been a mentor and hero to Stone and others. Still, Stone had complete confidence in his own diving skills and experience. And he was perhaps the only remaining expedition member, am Ende excepted, who still had complete confidence in his rebreather. So he knew that, by process of elimination, the one thing most likely to get him killed going forward with diving was an inexperienced, irresolute partner—am Ende, in other words.

It was very important not to let his emotions run away with him at this moment. Stone tried to sort things out rationally, factor by factor, but every scenario, every permutation, kept bringing him back to this: *We have come so far, to give up and go home now . . . just unacceptable.* His commitment was total. If he had not understood this before, he did now, with crystal clarity: for him, failing would be worse than dying.

He would go forward with or without am Ende. But for the safety of both, and for his own conscience, her own commitment had to be unshakable.

There was only one way to clarify that. He had to give her a clearly stated chance to opt out.

"Look," he said, "if you don't feel good about this, let's abort now."

Instead of relieving am Ende, the question angered her. She felt that Stone was doubting her, after all she had done to support the expedition and him. Staying with him after all the other expedition members—every one a man, to boot—had deserted him. Supporting him physically and emotionally through his personal ordeal in the desert of despair. Enduring the same privations and taking all the same risks for months. How could he do that, now? How *dare* he do that!

Stone hadn't meant to make am Ende mad. Quite the contrary, in fact; he'd only wanted to give her a clear, no-strings-attached opportunity, one last chance to say that no, this doesn't feel right, and we are taught that if dives don't feel right you shouldn't do them, and so I just can't do this one. His good intention had clearly misfired. But the jolt of anger actually helped, because it burned away am Ende's vacillation.

"Don't give me that crap," she snapped. "Let's do it. Let's put smiles on our faces and let's feel good about this."

It was May 1, 1994. Later that morning, am Ende slipped from the lower staging platform into the sump's cold, murky water and geared up with the help of Sloan, who had swallowed his objections when he'd seen that there would be no stopping the two. He was unhappy about the whole thing, no question, but he was too much of a friend, and too ethical an explorer, to abandon them down here.

She finned slowly around, making final checks of the rebreather and the backup regulators, settling her mind, waiting for Stone. He was soon in the water, laden not only with his 150-odd pounds of diving gear but also with the orange, 150-pound bag of food, carbide, and camping equipment that would sustain them on the other side if they found lots more cave to explore—"going cave."

Stone completed his own checks while Sloan watched from the lower staging platform. Finally ready, Stone reached up for Sloan's hand. "See you in a few days, brother," he said.

"Come back alive," Sloan said, gripping Stone's hand long and hard, and even the stoic Bill Stone was deeply touched by the love and concern he saw in Sloan's eyes. He turned to am Ende and took her hand. Then she held Sloan's.

"We *are* coming back," Stone promised.

She nodded, put the rebreather mouthpiece in place, and sank beneath the surface.

AM ENDE WENT FIRST, so that her more experienced partner could come behind, watch for problems, and help if they arose. She followed the white guideline that Broad and Rolland had strung during their dives. The visibility was less than 5 feet, and though Stone tried to keep her in sight through the water's blue-white haze, am Ende swam in and out of his vision like a ghost. Whenever she materialized, he focused on the buddy lights on the back of her rebreather.

Stay green, baby, he prayed silently, and they did, indicating that her rebreather was functioning properly.

They followed a narrow, steeply downsloping tunnel for roughly 450 feet before coming to the breakdown maze that had stymied the first dives in here just five weeks earlier. Both were able to worm their way through the small opening in the breakdown; beyond it the underwater terrain changed into a spacious canyon with a gravel floor, making it easier to avoid silt-up. Because using the smaller muscles of their arms consumed less breathing gas than the larger ones in their legs and lower body, they kept their legs still and pulled themselves along using rock handholds. About forty minutes after leaving Camp 5, they surfaced in what had been named the Rolland Airbell.

Several days of rain in mid-April had raised the water levels throughout the cave. Here in the Rolland Airbell, they were a foot above what Stone had encountered before. The top of the sandbar remained exposed, however, and when Stone and am Ende clambered up onto the beach, their lights revealed an eerie scene. Ian Rolland's footprints were still visible all the way to the sandbar's end, where, deposited by the flood, his empty boots floated.

I feel his presence, Stone thought.

They lingered here only long enough for an equipment check, then entered the second flooded tunnel, Sump 2, where the terrain was decidedly less benign. It dropped steeply, at almost 45 degrees, and sharp rock daggers hung from the ceiling. Gradually, the ceiling smoothed and the sump got so large that their diving lights did not show the bottom. Swimming slowly away from the airbell, they soon found the reel Ian had dropped more than a month earlier. Stone considered retrieving the reel but decided to leave it as a small monument to Ian's discoveries. They swam on.

All was silent except for the subtle hiss of gas moving through their rebreathers. Their wet suits held the chill at bay. In such an environment, the most active things were their minds and the biggest challenge was keeping them reined in. Some cave divers spend half an hour or more meditating before a serious dive, entering a quiet state of utter concentration they hope will be impervious to panic. Am Ende and Stone didn't meditate before dives, but they knew that focus meant survival and distraction invited disaster. Ironically, it was not am Ende but Stone who first came to grief.

TWENTY-ONE

ABOUT HALFWAY THROUGH THE TUNNEL, am Ende could not help stirring up silt as the sump's bottom rose. Stone lost sight of the guideline. He tried to surface to orient himself but found no airspace, only rock ceiling with water all the way up. Without a millimeter of free space, he was in a complete silt-out.

Stone knew standard procedure for locating a lost guideline. Cave divers carry a "gap reel"—a spare reel loaded with line. If they lose contact with the main guideline, they tie the gap line off to any available contact point and swim back and forth in widening arcs until they hit the main line. Tying off to that, they swim back to untie their gap line, then return to the main guideline, reeling up the gap line as they go.

But Stone could not find his gap reel. In fact, he had left it at Camp 5, a predicament that would have shoved many divers over the line from stress to panic. But the rebreather, which might have taken one life, now saved another. Stone's large body needed a lot of air. Had he been using conventional scuba tanks, especially with accelerating heart rate and respiration, he might well have run out of air too soon. But his rebreather gave him *hours* of dive time, making all the difference. *Relax*, he told himself. *You've got time to think this through.*

He methodically worked out a plan that would make good use of a God-given gift: his six-foot, four-inch height. If he spread his arms and legs wide, the additional twenty-four inches of his fins would create a fingertip-to-fintip span of more than ten feet. Deflating his buoyancy vest, he floated down in a skydiver's spread-eagle position and hit the guideline with a fin on his first try. Grabbing hold, he followed the line to the end of the sump. Am Ende was already there, waiting for him on a gravel beach.

Not a good time to freak her out, he thought. He took her hand and said, "Very nicely done"; he made no mention of his narrow escape.

They established Camp 6, where they would spend their first night, on a gravelly hill 300 feet beyond the sump. ("Camp" was mattresses made by putting their two wet suits into plastic garbage bags.) Then they confronted an excruciating decision: what to do with the rebreathers, their sole hope of escape from the cave. Having dived their way through the two sumps, there was no way they could free-dive their way back. Their survival depended on functioning rebreathers.

Am Ende's rebreather had already suffered that one malfunction during Rolland's dive on March 25, the day before his death, when it exhausted its carbon dioxide scrubber prematurely. That failure, plus a punctured bladder on Bill's unit and the two problems with am Ende's at Camp 5, had amply demonstrated that these may have been marvels of invention, but with hundreds of delicate parts and space-age electronics, they were far from invulnerable.

The fact was that every time they moved the rebreathers, powered them up and shut them down, or submerged and resurfaced, they increased the probability of malfunction. Beyond their camp, they saw a number of dry passages, so diving was not their only option for going forward, at least initially. They finally decided that they would do no further diving on the rebreathers until it was time to return through the sumps, both to protect them from damage and to leave them with the maximum reserves of breathing gas. After powering them down, they cached them on a rock outcropping as high as they could, hoping that if—no, *when*—the rains of summer began, whatever flow resulted would not be high enough to carry the rebreathers away. Of course, if flow that high came roaring through, it would wash both of them away like corks in a sewer main as well.

The rains of summer. They were now playing a Mexican cave version of Russian roulette with those rains, which were notoriously fickle. The rains might come in weeks, or days, or even hours. No one could say for sure. But when they did arrive, the cave would flood. And it was easy for them to envision how they would die if that happened, because something similar already had.

Two weeks earlier, on April 16, a rainstorm had struck. It lasted just three days, but this early storm gave a preview of what floods could do, raising river levels in surface canyons six feet, sending torrents roaring down into Huautla. In the cave, meandering streams became whitewater rivers, and dry vertical pitches morphed into lethal waterfalls.

The floods trapped a *National Geographic* photo team and a few others at Camp 3. Most were seasoned explorers, inured to the trials of supercaving. Still, after two days and nights of confinement, it became hard not to envision a wall of water blasting through the great chamber, washing them all down into the cave like bugs in a toilet. After three days and nights, the imprisonment became too much and they bolted for the surface. It was a bad decision. During the attempt, Steve Porter nearly drowned twice before they had gone a half mile, at which point everyone hustled back to Camp 3, wetter but wiser. Eventually all returned safely to the surface, but the cave had made its point. Am Ende and Stone's chances of escaping after significant rainfall would be slimmer, because they were so much deeper and would feel the full force of the endless rainy season downpours.

Am Ende and Stone knew about what had happened to the group at Camp 3. They also knew that the beginning of rainy season fluctuated from year to year. They understood, as well, that they were at the bottom of a 6-mile-long drainage system that would collect water from the entire surface region and funnel it right toward them. If it rained while they were underground, it would be a fatal disaster.

As it turned out, their first disaster was caused not by rain but by light. Or, rather, by its absence. Standing atop a boulder in an immense breakdown pile just beyond Camp 6, am Ende took off her helmet to tighten the headband. Her carbide lamp fell out of its helmet mount and dropped down between the boulders at her feet. Light was a finite resource here, more important than water or food.

Stone was in another area, drawing in his notebook. "We have a problem here," she called. He came over and she explained.

At first Stone didn't believe her: "You're pulling my leg."

Am Ende assured him she was not. He peered down at the pile of rocks. "Down there? Can you see it?"

She shook her head. "Maybe you can reach it . . ."

Stone gave her his helmet, which had its own carbide lamp attached, and flopped down onto the pile of boulders. The spaces between them were small, but his lifetime of worming and squirming that six-four frame through viselike squeezes helped. Forcing his face and shoulders down into one crack, he worked his right hand, with a flashlight, into another. The brass lamp shone in his flashlight beam, 10 feet deeper. It was perched right at the lip of a small ledge. One clumsy brush would send it clattering down into the depths, lost forever. He stretched to his limit, but his fingertips were still at least 5 feet short of the lamp. "I can see it, but I can't reach it," he reported.

He pulled himself out of the cracks. Just as he did so, he heard am Ende exclaim, "Damn, not again!"

"What's wrong? What are you doing?"

She had been standing by in the dark, flashlight off to conserve batteries. "I was drying out the felt in your lamp," she said miserably. Never the useless bystander, am Ende had taken Stone's own carbide lamp apart to dry out its wick. As she worked by feel in the dark, saving those precious batteries, the unthinkable had happened. She'd dropped Stone's igniting mechanism, an assortment of tiny parts—flint, wheel, striker, spring—that they would never, ever find again. The first lamp loss had reduced their primary light source by a third. She had just cut it by another third. Down here, except for air, light was their most precious life-preserving resource. This was a crisis that threatened both their lives.

TWENTY-TWO

THEY EACH HAD TWO ELECTRIC RESERVE lights. With brand-new batteries, each one of those would give about three hours of light. But the batteries were not brand-new, and Stone had no way of knowing how much juice any of them had left. They did have three other electric lights each, but those were diving lights and could not be used for *anything* but their return trip through the sumps. Finally, they had one spare carbide lamp with a week's worth of fuel and an intact lighter.

If they went several days deeper into the cave and anything happened to that last carbide lamp, they would not have enough light to regain Camp 6 and still make the dive through both sumps. They would be stranded out there beyond all help in the absolute dark, where they would die in one of three ways. They might die slowly and horribly from starvation. Injury might incapacitate one or both of them, in which case they would still die slowly and horribly, just with more pain. Finally, with no hope of escape, they might opt for quicker but more brutal deaths by taking their own lives—if they could figure out a way to do that in the dark.

"I'm sorry. I screwed up," am Ende said miserably. "I didn't do it on purpose." She was mortified. The first drop had been a fluke, but she knew that the second had resulted from bad decision making. If she had paused to weigh carefully the possible loss of their second carbide lamp against the use of a few minutes' battery power, she would have opted to use the batteries and safeguard the carbide lamp.

Stone *was* angry now. He sat beside her, icily silent. It was hard not to lash out, but then he heard Marion Smith's voice in his head, laconically shrugging off that near-fatal canteen drop in Fantastic.

"It happens," he said to am Ende. The tension evaporated.

But their crisis did not. If anything happened to that last carbide lamp, everything they had sacrificed up to now would be for naught. Stone thought, *Barb's lamp is lost. Mine is unsalvageable. We can't go on with one carbide lamp. No choice. We find her lamp or we retreat.*

How to find her lamp? They could not reach it. They had no hammer drill or explosives to break up obstructing boulders. They could not tunnel under the rock pile, because it was far too deep at this point. They had no tools long enough to retrieve the delicately balanced lamp. They could not even jury-rig a replacement, Apollo 13–style, because they did not have the parts.

"We're gonna have to move some rock," Stone announced.

It was the only option, but at first blush a poor one. This entire pile of boulders—the whole cave, in fact—was composed of limestone, which weighs 163 pounds per cubic foot. A block no bigger than a desktop computer tower weighed almost two hundred pounds. A boulder the size of a washing machine weighed more than one thousand pounds. Most of the boulders in this pile were that big or bigger.

If Stone ever needed a Doc Savage solution, this was the time. One of the great benefits of engineering was that it had trained him to focus on options rather than obstacles and to think in terms of systems. Now his challenge was to create a system that allowed both him and am Ende to apply the maximum achievable force simultaneously and with optimum efficiency.

It happened almost automatically. He quickly decided that a push-pull combination would be most effective. He would be the puller, and he devised a connection that allowed him to apply all of his two hundred pounds in the most dynamic way. He tied a loop in a length of tough, orange nylon webbing and used it to "lasso" the topmost boulder.

Am Ende would be the pusher. She braced her back against an adjacent boulder and pressed her feet against the one wearing Bill's sling. This position engaged her body's strongest muscles—abs, glutes, quads. At his signal, he pulled and she pushed with her long, powerful legs. The boulder rocked, hesitated, passed its tipping point, and crashed down into darkness. They worked for a long time, moving boulders that were within Stone's leg-and-lasso system's capacity. That left the lamp still several feet beyond his reach.

He had thought this far and was ready. Knotting together odds and ends of parachute cord, he fashioned a "fishing line." He had no hook, and the lamp's position made it impossible to snare it with a loop, so he tied a small knot at the end of his line.

Maybe I can wedge it somewhere or snag a corner, he thought. He lay down on the rocks again, with his face pressed into a new crack, flashlight in his left hand, "fishing line" in his right. Keeping his flashlight's spot focused on the lamp with one hand, he "unreeled" the line with the fingers of his other hand. There were several places on the lamp where the knot might catch, if only he could position it properly. But he had to be exquisitely careful to avoid nudging the lamp off its precarious perch and losing it for good.

He worked for forty-five minutes in that contorted position, reeling and unreeling, swinging the line delicately back and forth, striving to gain purchase. His muscles burned, and he sweated like a laborer despite the cave's chill. Am Ende could only sit helplessly off to one side in the dark, suffering agonies of guilt. More time passed. Both began to despair.

Stone knew there was nothing to do but keep trying. He had finally zeroed in on what he thought was the best spot to snag, a tiny slot between the bottom of the lamp's circular chrome reflector and the top of its cylindrical brass case. That space was just smaller than the knot. If he could get the knot behind the slot, he might be able to pull up ever so gently and lift the lamp within reach. If the line had been rigid, a stick, say, he could have pushed the knot back behind the slot fairly easily. But a rope won't push. He could make the placement only by a perfect little swing of the line and knot.

Finally, after hundreds of failed tries, he managed to nudge the knot behind the slot. He gathered in the line as though it were spider silk, knowing that if he dropped the lamp, all was lost. Slowly, slowly it approached his reaching fingers. And then he had it.

He jumped up and held the lamp high over his head. Am Ende let out a shout of relief and joy. She tried to hug him, but he was still angry, not quite ready for hugs and high fives. He waved her off. "Tell you what," he said, in his best deadpan, Marion Smith voice. "Let's tether these things to our helmets from now on."

They celebrated that night with a big feast of freeze-dried beef Stroganoff. By that time, Stone's anger had cooled and his ardor had warmed. Their celebration included some postprandial activity which guaranteed that even if the cave did not set a record for depth, they'd established one of their own in Camp 6.

TWENTY-THREE

THE NEXT MORNING, MAY 2, after a seven o'clock oatmeal breakfast, they loaded up with ropes and climbing hardware and headed deeper into the cave. Before long, their path was blocked by yet another huge, impassable sump filled with dark water. Diving it was not an option. Looking for some other way to get past the sump, Stone searched the surrounding walls with one of his electric lights. Down at the sump's far end, about 25 feet above the water, was a door-sized opening. Like someone tiptoeing along a second-story windowsill, Stone inched along a long, narrow ledge that ran around the cave wall until he was beneath the opening. Then he free-climbed straight up the wall, hoisted himself into the opening, and rested for a moment. Sitting there, he saw a Snickers candy bar wrapper wedged into a crack even higher on the wall than he was. It could only have been washed there by the earlier, mid-April flood. It was sobering to think that just three days of rain had raised the water level here more than 25 feet. Rainy season flooding would be much, much worse. But there was nothing he could do about that. Returning to the task at hand, he set a bolt to make following easier for am Ende, who was carrying the haul bag with their ropes, hardware, water, and trail food.

Continuing through a series of passages and tunnels, they found themselves above another giant lake. This time, though, almost directly beneath them, a pyramid-shaped rock rose 40 feet above the water. They climbed down onto the rock, which was surrounded on all sides by water of the largest sump they'd yet encountered. It was as if the deeper they went, the more the cave worked to frustrate them, like that moment in a novel or movie, the climax, when a hero is finally confronted with the Final Obstacle, something so big and dangerous that she cannot possibly surmount it (but somehow always manages to).

What enabled them to keep pushing on, fully aware that every additional moment spent down here, separated from all possibility of rescue, relegated them to the position of long-serving combat soldiers who know, beyond all doubt, that for them it is not a matter of whether, but only when, catastrophe will strike?

It was partly that they were used to being in the cage with the lions. They had been in caves before—other supercaves in Stone's case and, if not those, at least other very big ones in am Ende's. Also, they had gone into this with their eyes wide open, having considered the risks, weighed them against the benefits, and found the outcome acceptable. They had, in other words, made peace with the possibility of their own deaths. None of this is to suggest that they were not afraid. They were. Absence of fear would have been a sign of mental imbalance. But as so many others have observed, true courage is not the absence of fear; it is the ability to persevere despite it. And persevere they did. They had felt the breath of the lion so often that by then it was almost like a kiss.

There was only one way to find out what lay beyond, so Stone started swimming. He was wearing a fleece jumpsuit but no wet suit, and the water was cold—64 degrees, 34 degrees lower than his body temperature; water steals body heat seven times faster than air. He explored three corners of the sump, finding blank walls each time.

By the time he started his fourth and final probe, to the southwest, Stone was hypothermic. He could make out a sandbar about 100 feet across the water. But when he reached the midway point, he realized he was going to drown. His soaked clothing, the weight of his vertical gear, and his flooded rubber boots were all pulling him under. And he could not touch bottom.

He tried to find the bottom with his toes, but could not. He flailed with his arms, feeling the sharp edge of panic. Hyperventilating, he felt the water start to close over his mouth and nose. His head went under.

At that moment, his toes touched the bottom. Panting, spitting water, he clambered up the sloping shore. After a few minutes, he was able to yell to am Ende that he was okay. Carrying less weight, she followed. Stone climbed up onto the bank, at the end of which he found a passage 50 feet in diameter and with an even surface that he could take at a run. After the length of a football field, he stopped and stared. The already sizable tunnel flared out dramatically and kept on going. It felt to him as though he had just stepped onto the far side of Pluto, a place more remote, intimidating—and exhilarating—than any he had ever been.

This was almost surely the magic portal at last, the passage that would lead all the way down to that long-sought resurgence in the Santo Domingo River where Jim Smith's dye had turned the water bright green. If it went, as he felt sure it would, Huautla would become the deepest cave on earth by a wide margin and he would finally have proof that the last eighteen years of his life had not been one great windmill tilt. He jumped and shouted like some ancient pagan worshipper, and the cave rang with his howls.

Beyond the portal they discovered one of those Jules Vernesian features supercaves offer up from time to time. Seven miles from the cave entrance, almost a vertical mile beneath the surface, they entered a vast, teardrop-shaped chamber 500 feet long, 450 feet wide at its broadest expanse, and 100 feet high at the top of its arched ceiling. Naked numbers can't begin to suggest what such a place is really like. The floor alone of Perseverance Hall, as they named the chamber, could have contained more than fifty of those giant diesel locomotives Stone had imagined hearing back in Camp 5. If they were stacked in two levels, which the ceiling height allowed, it could have held twice that number.

Stone and am Ende were ecstatic. Only vast flows of water over eons could have carved out such a monstrous room, and that water must have kept on flowing beyond, presaging even more incredible finds. Boulders covered the floor of Perseverance Hall, which sloped down at about the angle of a home staircase. Here, finally, excitement got the better of them and they started to hurry. Am Ende went down first, with Stone following. Their rush was understandable, but in caves haste often makes death. Stone stepped on a sofa-sized boulder that looked stable but suddenly began sliding, then rolling downhill. They later estimated that it weighed ten thousand pounds. He struggled to stay on top, like a lumberjack running on a spinning log floating in the water. Am Ende heard the crashing, shouted, got no answer. Down there, disaster could strike within hailing distance but remain invisible.

Stone launched himself off the runaway boulder and slammed to a stop with his back against other rocks, facing uphill, watching the limestone giant roll toward him. It happened in microseconds, so he was like someone who steps out in front of an onrushing bus, sees it, understands that it's going to hit, but doesn't have enough time to get out of the way. And then, miraculously, the boulder just . . . stopped. It was caught and held by another rock formation, a giant marble dropping into a slot.

Am Ende found him like that, sitting against a rock, staring in stunned amazement at the massive hunk of stone that had come within milliseconds of crushing him to death. Then he got up and out of its way—fast.

They both understood that if Stone had been injured, am Ende would have had to leave him, retrieve her rebreather from its high cache, dive alone back through the two sumps, and hope that Noel Sloan was still there at Camp 5. With Noel or alone, she would have to spend two days climbing all the way out of the cave and then somehow summon help. But help from *where*? The Mexican government could not perform such a rescue. Other cavers, presuming they were willing and able to attempt a rescue (or recovery), could take weeks to reach Stone.

Meanwhile, Stone would be as alone as a man on the moon. Open wounds would quickly become infected in the cave's dark, damp, microbe-rich environment. Fractures could kill in other ways, a fat embolism being one. Exposed by a break, bone marrow may leak microdrops of fat that, once in the bloodstream, can cause fatal damage to the lungs and brain.

They had had enough for that day. A good night's sleep at Camp 6 calmed their rattled nerves, and the next morning, May 3, they were ready to attack again. They settled on a simple plan: go until, for whatever reason, they could go no more. This time they wore wet suits. They packed ten pounds of carbide, all their rope and climbing gear, and what remained of their supply of gorp, a high-energy mixture of raisins, candy, and nuts.

Moving quickly, in three hours they passed through Perseverance Hall and, at its far end, arrived at the edge of another big sump. This time it was am Ende's turn to get wet first. With a rope tied to her harness, she swam to the sump's far end and then disappeared into a crack. She had only about 12 inches of airspace between the water's surface and the crack's ceiling, which got lower the farther in she went.

If the water rose, she would have to push her nose or lips into that airspace and breathe that way. Cavers will do that, struggling to breathe in clear space the width of a pencil. But the weight of a helmet and carbide lamp can quickly tire the neck muscles, making it harder to keep the head in a position that allows breathing. If one is far from such a crack's entrance or exit, the muscles can suddenly cramp or give way. Sucking in a lungful of water can precipitate panic. And panic usually has only one outcome down so deep.

Stone watched am Ende disappear into the crack, trailing rope. For a while he could hear her caving helmet scraping against the crack's ceiling, but then that stopped, and the glow of her lamp disappeared. He was left alone in the dark, slowly paying out line. Recalling his own experience at the end of a rope, he wondered how long he should wait before trying to reel her back. Ten minutes passed, the rope still going out slowly, then fifteen minutes, and then all movement stopped. Was she in trouble? Or had she stopped to rest, or adjust some piece of gear, or look at something? He had no way of knowing, and now he understood how agonizing it must have been for the two men on the other end of his line during that earlier dive of his.

More minutes dragged by. Reel her in? But what if she was in an awkward position where a sudden yank could injure or drown her? No, he would have to wait for some kind of signal.

He looked at his watch. *Twenty minutes.* Because he did not want to pull on the rope without knowing her situation in the crack, his only alternative was to go after her himself. Struggling to contain the horrible thought that he could be dealing with yet another death—this time of the woman he loved—Stone slid into the water.

TWENTY-FOUR

HE WAS CHEST-DEEP IN THE frigid water when he suddenly felt three sharp tugs. Though years to come would bring other close calls in caves, he would never again feel relief so intense. Rather than haul her back in a panic, as his own handlers had done, nearly bringing him to grief, he simply held steady on the rope so that she could bring herself back. Am Ende finally reappeared, cold and dripping but excited.

"It goes," she announced. "I used all the rope. The swim is at least sixty meters. And there is a big river tunnel on the other side."

That excited Stone, too, but the tight 200-foot passage that am Ende had just traversed made him uneasy. They had no way of judging the oncoming, or even current, weather. He knew that even light rains had sealed cavers on the far side of such passages before. Those victims had been saved either when the rains stopped and the water levels fell or by divers who'd rescued them. (In some cases, the divers had performed recoveries rather than rescues.) But if that happened here, as deep as they were in Huautla, there would be no rescue. And because it was now the rainy season, the water level would not drop—it would keep rising for weeks or even months, in fact.

There was nothing they could do to stop the rain, but Stone decided to at least rig a rope through the passage, thinking that they might be able to pull themselves back if it flooded. The security that rope offered was mostly psychological, because a rainy season flood that trapped them on the other side would almost certainly kill them. For one thing, without diving gear of any kind, they would have to come back through the flooded tunnel holding their breath. Making a cold, 200-foot breath-hold dive encumbered by packs and gear, even in open water, would be a small miracle in itself. But there was more. If am Ende got partway through and decided to abort, could she make it all the way back before her body's autonomous carbon dioxide reflex made her gasp, drowning her? Coming second, Stone would have no way of knowing whether she had made it. (Pulling along a second rope behind her, as Stone had done back in '79, would only increase her drag and the possibility of snagging; the margin for error was too thin here for that.) After waiting some reasonable amount of time, he would start through the passage, perhaps only to find her dead body blocking the way forward.

Nor was that all. A well-known phenomenon in physics called the Venturi effect dictates that when fluid flows from a larger area into a smaller, its velocity increases. Garden-hose spray guns work because they constrict the slower flow from the hose through the gun's tiny nozzle. If enough water flowed down through Huautla to fill the preceding sump and force it through the smaller passage beyond, it would not be like a garden hose. Water would blast out of the passage's far side as if from an open fire hydrant. Nevertheless, am Ende explored on by herself while Stone worked, and came back just as he was finishing the second bolt.

"Do you want the good news or the bad?" she asked.

"Give me the bad first."

"I followed the water downstream past another lake. It turns into a sump. I swam the whole perimeter. There's no way on."

"So what's the good news?"

She explained breathlessly that about 300 feet ahead of them, she had found a huge river pouring in from the left side of the cave with four times the flow of the one they had been following.

"Four times? That's incredible."

"Yes, but it leads upstream. That's not the direction we want to go." Unable to conceal her disappointment, she said, "I think we're at the end of the road."

"We'll see about that," Stone said.

It took them only about fifteen minutes of wading through shallow water to find the waterfall she had described. Stone, too, was amazed. It was the biggest waterfall he had ever seen underground. Water was jetting straight out of a hole in the rock that was easily 20 feet in diameter. Stone believed this was, at last, the mythical underground resurgence of a surface river, the Río Iglesia, for which two dozen expeditions had searched since 1967.

They pushed on, moving through a series of narrow, ascending rock passages that eventually deposited them in a monstrous room. It was about 330 feet in diameter where they entered, then narrowed like a funnel, down a 45-degree sand slope, to yet another sump, this one 165 feet wide and 80 feet long. It crossed the end of the chamber like the top bar of a T. They had hit the sump perpendicular to its long axis and about at its midpoint. The wall on the far side of the water was sheer and rose beyond the limits of their lights. Into this sump flowed all the water from all the waterfalls and rivers and channels and sumps and infeeders above them. There was only one name for it: the Mother of All Sumps.

Large transverse dunes crossed the 45-degree sand slope like giant steps, and they knew that only a truly monstrous flow (visualize Class IV rapids) could have sculpted sand, gravel, and rocks into ridges like that. Every rainy season produced just such torrents, which came roaring and boiling down the riverbed, carving its bottom and dumping everything into the Mother Sump. Am Ende and Stone were standing at the end of the barrel of an immensely long gun, one the rainy season could load and fire at any moment.

There was no way around the sump, no way through it or under or over it.

"Checkmate," Stone said aloud.

STONE AND AM ENDE RETURNED TO Camp 5 without incident, then continued on to Camp 3 and out of the cave. Their accomplishment had no parallel in the annals of cave exploration—and few in exploration history, period. Years later, Stone was asked by a NationalGeographic.com interviewer to name the happiest moment of his life. His quick answer said worlds about what he and am Ende had endured during their extraordinary six days beyond the sump:

"The evening of May 6, 1994, toward the end of the four-and-a-half-month San Agustín Expedition. That was when, after 11 days underground in the Sistema Huautla caves in Mexico, my colleague Barb am Ende and I managed to make it back to Camp 3." Their six-day Huautla effort was a monumentally stunning accomplishment, and yet Stone's happiest moment came not during their exploration of the great cave but upon their escape from its darkest heart.

Stone had been seriously exploring the caves of Mexico for eighteen years, since his first true expedition in 1976 with Jim Smith. He had been on this expedition for almost three months and deep below ground on the last push for eleven straight days and nights. Based on measurements they made during their six-day foray, he and am Ende had established that Huautla, at 4,839 feet, was the deepest cave in North America. They had explored more than 2 *miles* of new passage, the entire time exposed to countless dangers and beyond any hope of rescue.

Their discoveries gave am Ende and Stone cause for rejoicing. But the ultimate goal still eluded him. The deepest cave in the world, just then, was a French cave named Réseau Jean Bernard, which, at 5,126 feet, was 287 feet deeper than Huautla. Thus ended 1994.

Well, not quite. Shortly after Stone and am Ende walked out of the cave, *Outside* writer Craig Vetter walked into the camp.

TWENTY-FIVE

"THERE'S ONLY ONE PERSON I'D BE more surprised to see down here, and I'd have to kill that person," Stone said when Vetter appeared in early May. (The homicide reference was to an absent filmmaker.) Stone was a bit strung out. The sting of *Outside*'s 1992 piece—at least that sharp ending—might have not entirely faded yet, either. And it appeared that *Outside* had sent a writer down only after its editors had heard juicy rumors of "bitter dissension and death," as the article would later say. Apparently not much interested in a successful expedition, the magazine seemed quite interested in one beset by disaster, death, and desertion.

The expedition was essentially over by the time Vetter arrived. The remaining team members were emaciated, drained physically and emotionally, "a picture of dirt-eating exhaustion." The atmosphere was rank with anger, disappointment, and grief. There had been a death, a mutiny, countless close calls, and many vicious arguments. The leader's woman had been more than a small part of the problem. All in all, it looked like the ultimate expedition nightmare.

Which made it a magazine's dream.

Outside had stumbled onto a veritable bonanza of scandal, death, and intrigue, the kind of thing that might come along once a decade. The resulting article, "Bill Stone in the Abyss," took full advantage. It ran more than seven thousand words at a time when three thousand words was a long standard feature. Its subtitle:

> **His life's obsession has been to get to the bottom of the world's deepest cave. Two team members have already died. How much farther is he prepared to go?**

The article referred to Stone as a sorehead who was obsessed, suicidal, sullen, surly, possessed, crazy, callous, desperate, pompous, and hyperinsensitive. It asserted that in his zeal to establish Huautla as the world's deepest cave, Stone would "choose death before outright failure."

Its conclusion declared, "In the end, it may be that prizes like Huautla go only to those who rarely ask anyone's permission for anything, and who rarely stop to count the price."

There it was again, another of those shape-shifting bits of writing that tiptoe around the edge of libel, seemingly saying two things at once, or maybe nothing at all. To be fair, the article did include a few positive references, but, buried in that avalanche of seven thousand mostly critical words, they were hard to find.

Stone was furious when the piece came out several months later. He thought it a hatchet job by a publication willing to stoop to any level to sell copies and, to this day, he calls the magazine *Outhouse*. Craig Vetter, to this day, defends the article as fair and balanced, saying that he reported only what he found. His claim is credible, given that *Outside*'s assignment came late, so that by the time Vetter arrived, he could only poke through the expedition's charred bones. But how to account for the overwhelming number of criticisms in the piece and so few direct responses from Stone? "He was difficult to interview, and Barbara am Ende was his gatekeeper and guard dog," Vetter told me. Partly, Stone's reticence grew out of his displeasure with Vetter's unexpected arrival. More compelling, though, was his contract with sponsor *National Geographic*, which forbade him (and the other expedition members) from speaking with any other media. That fact did not find its way into the *Outside* article.

Be that as it may, it's safe to assume that most of *Outside*'s million-odd readers did not come away from the article impressed by Stone's solo recovery of Rolland's body, which got one sentence, or by his and am Ende's incredible six-day exploration (several paragraphs), or by the rebreathers' flawless performance. Nor would most readers have come away thinking of Bill Stone as one who, like Livingstone, Shackleton, and Lewis and Clark, overcame seemingly insurmountable obstacles in the name of exploration and in pursuit of a great discovery. Most likely they would have ended up thinking of Bill Stone in very different terms—as, say, the ogre-in-chief.

But set all that aside for the time being. Magazines publish what sells, and we buy what they publish. The more compelling question, looking back over the whole Huautla episode, is: *Why?* Why push himself and others to such extremes, and why at that particular time? On previous expeditions, like those to Peña Colorada, Stone had been disciplined but not overbearing. (Not overbearing enough, perhaps, to judge from the Peña Colorada mutiny.) At Huautla in 1994, "overbearing" was probably too mild a descriptor. Something, apparently, had happened to Bill Stone between Peña Colorada and Huautla. What was it?

One thing is obvious: he was older and feeling pressured by the advancing years. In Vetter's article, one expedition member said, "Through it all, Stone pushed as if he were late for something." Type-A personalities *always* feel as if they are late for something—it's the condition's defining characteristic. Moreover, Stone really *was* late for something. In 1994, he was forty-two years old. For a lawyer or teacher or bus driver, say, that would have meant nothing. But for those who make a living with their bodies, like professional athletes, models, prostitutes—and explorers—forty-two is the leading edge of old age.

Caves are no country for old men. During extended expeditions, their assaults on the body are cruel and numerous: weeks of rappelling down and climbing up immense vertical drops with huge loads, banging like human wrecking balls into rock faces, scraping through rib-cracking squeezes, destroying knees on steep breakdown piles, worming through spine-twisting breakdown, all the while soaked and verging on hypothermia, buried in perpetual darkness, malnourished, sleep-deprived, and diarrheic. (All this in addition to cave diving, of which Stone was a frequent and extreme practitioner.) At forty-two, he had been taking that kind of beating annually, or sometimes more often, for twenty-two years, and he was already older than many others who were going on, let alone leading, supercave expeditions.

Perhaps he also was feeling pressured by the greater success of other cavers, many of them younger, elsewhere. European cavers had been doing impressive work on their continent, particularly in France and Austria, where supercaves still being explored had been swapping the "world's deepest" record of late. And rumors had been coming out of eastern Europe—the Republic of Georgia, especially—about a couple of supercaves being investigated by expert, well-equipped, highly organized international expeditions.

It might also have been money. Stone routinely invested his own funds, without reservation, in his expeditions, and he had gone deeply into debt to launch the 1994 effort. His above-ground lifestyle reflected, if not impoverishment, at least a Spartan quality. He lived alone in a nondescript, sparsely furnished home in a Washington, D.C., suburb where Rabbit Angstrom would have fit right in. The house, in fact, was not really a home in the traditional sense; it was more like a way station between caverns. *Caves* were his home and, to a lesser extent, the laboratory.

Living without independent means meant finding sponsors. The corporate world was watching, not only those who had already ponied up but everybody who might conceivably back him in the future. Much of the time during Huautla '94 he felt something that no one else did: the hot breath of those sponsors on his neck, peering over his shoulder at every turn, interested in only one thing: ROI, return on investment. We showed you the money. Where's our payoff?

But does even all that explain Stone's pushing himself and others to such extremes, risking so much, including his lover's life, and suffering such costs, including Ian Rolland's death? One quick and easy answer pops to mind: obsession. In fact, some 1994 expedition members did use that word.

Obsession has fascinated us since biblical times, Jesus being perhaps the ultimate example of obsession's triumph—and toll. The word conjures visions, most wildly pejorative, of characters like Mr. Kurtz, the deranged ivory trader in Joseph Conrad's novella *Heart of Darkness*. (*Outside*'s article did suggest, obliquely, that Stone might have become a real-life Kurtz.) In the minds of many, "obsession" is followed quickly by "crazy."

Obsession is not classified as a mental illness. Its dictionary definition is "domination of one's thoughts or feelings by a persistent idea, image, or desire." So one doesn't have to be crazy to be obsessed. But like the rich, the obsessed surely are different from you and me. Edward Bulwer-Lytton observed tellingly that "talent does what it can, but genius does what it must." Not all geniuses are obsessed, of course, any more than all the obsessed are geniuses, but the two conditions do share a certain lack of volition. One could even paraphrase Bulwer-Lytton to say that interest does what it can, but obsession does what it must.

Bill Stone himself dislikes the word, as did another great explorer, George Leigh Mallory. The pioneering Everest climber (he of "Because it's there" fame) was a husband and a father, like Stone, of three children. But the center of Mallory's short adult life was unquestionably Mount Everest. He made three attempts on the mountain, and the last, in June 1924, during which he may or may not have reached the summit, killed him. He left a wife, Ruth, and the children.

Robert Macfarlane, a gifted writer and mountaineer, comes to a startling conclusion about Mallory in his book *Mountains of the Mind.* In an earlier book of my own I quoted this same passage, and I cite it again here because it illuminates so brightly the subject at hand:

> To read Mallory's letters and journals from the three Everest expeditions . . . is to eavesdrop on a burgeoning love affair—a love affair with a mountain. It was a deeply selfish love affair, which Mallory could and should have broken off, but which instead destroyed the lives of his wife and children—as well as his own.

Not obsession, then, but—who'd have thought it?—*love.* Hmm. At first blush, that seems a middling excuse for megalomania. But not so fast. Another great writer on the subject of love pointed out that

> When a man loves a woman . . .
> He'll trade the world
> For the good thing he's found

He'll trade the world . . .

Obsession may be part of love, and love part of obsession. What is perfume about, after all, but love, and what is one of the most popular *essences d'amour* called but Obsession? If more proof of love and obsession's link be needed, consider this: Obsession's elegant bottle resembles nothing so much as the male organ, ready for love. The sharpest marketers on earth, who know a lot, understand that the two conditions are as intertwined as lovers on a bed.

More proof may not be needed, but it does exist, provided by none other than Bill Stone himself in a 1994 *Washington Post Magazine* article, "Journey Toward the Center of the Earth," by Hampton and Anne Sides. About his extreme caving, Stone said, "It's been a very insidious involvement . . . one of those things I absolutely must do."

There was undoubtedly another factor, and it harked all the way back to Stone's competitive father, Curt. At every opportunity, Bill Stone downplays the idea of competition in what he does, but that's a bit disingenuous. He is one of the most naturally competitive people you will ever meet, for one thing, and for another, competition has always driven those seeking the Big Discoveries. There are too few of them, and too many humans, for such prizes not to become hotly contested. Examples abound in every field. The fact that Scott and Amundsen, to cite them one last time, were literally racing to the South Pole on different routes is one example. Other scientists were hot on the heels of Crick and Watson, who discovered DNA, or, as they put it, "the secret of life." If Charles Lindbergh had not made his transatlantic flight when he did, others were idling on the tarmac, poised for their own takeoff rolls.

Exploration, be it for science or adventure, and competition go hand in hand. In 2006, Stone told *X-Ray Mag*, an international diving publication, "The *game* was that we were trying to *beat* the French. . . . Now we are trying to *beat* the Russians but it is still the very same *game*." (Italics added.) He could have been talking about the Olympics.

UNDERLYING EVERYTHING ELSE MAY HAVE BEEN something even more primal and basic: desperation. By 1994, Bill Stone had lost his wife, his children, his home, two close friends, and all his money; he had borrowed heavily and promised the moon to scores of sponsors. One golden moment had been snatched away by fate in Cheve in 1991. Stone knew that life did not often present a second, but there it was, down in Huautla, in 1994.

"I've shot my wad," he confessed miserably to Noel Sloan after dinner on April 29 down in Camp 3. "This is my chance, maybe my only chance. I'm going to push this cave to the bitter end. No matter what."

No matter what.

Bill Stone may have despised *Outside*, but there was an eerie ring to those three words. The magazine's 1992 article about him ended with the same three.

TWENTY-SIX

STONE-LED EXPEDITIONS IN 1995 and 1997 failed to find any other way through Huautla. Stone devoted 1998 largely to a massive project mapping Florida's vast Wakulla Springs underwater cave system. The project provided him with a matchless opportunity to test and refine his rebreathers. It also afforded the media yet another chance to have at him. This time it came in the form of a writer from *National Geographic Adventure*, the offspring of *National Geographic* magazine, published by the National Geographic Society, which was one of Stone's most important sponsors. Though he might have wished to, given his previous experience, there was no way Stone could keep NGA's writer, Geoffrey Norman, away from the springs.

The Wakulla Springs Mapping Project was three months long; involved a team of 152, supported by 28 sponsors; and cost more than $1 million. It was also ambitious, even by the standards of Stone endeavors, with not one but four goals. Improving and testing the rebreathers was one. Demonstrating the worth of another Stone invention, a computerized, sonar-enabled device that produced three-dimensional, color underwater maps, was another. The third was demonstrating the utility of a "variable depth decompression habitat," as Stone called it, a bell-shaped, submerged chamber that allowed divers to do long decompression stops in its warm, dry environment, rather than spending hours hanging off hand lines underwater. And there was the overarching reason for being there: mapping Wakulla's caves.

Early on, one of Stone's divers suffered a bout of decompression sickness (the bends) but survived. Later, another went into convulsions at 100 feet (oxygen toxicity, again) and nearly drowned. That diver was *very* lucky indeed to live. On February 15, a Nobel Prize–winning physicist and experienced diver named Henry Kendall, who had been working in support, did drown. All three had been using Stone's rebreathers; the first two accidents occurred because the divers had failed to adjust their units correctly. An autopsy ultimately revealed that a stomach hemorrhage killed Kendall. "He would have died in Wal-Mart," a Florida coroner was quoted as saying in Norman's article.

These mishaps would have been bad enough by themselves, but Stone and his team also ran afoul of territorial local divers who had been mapping the springs on their own for ten years. The 160-diver group called itself the Woodville Karst Plain Project (WKPP). Given their professed commitment to science, WKPP's divers might have welcomed the newcomers and, with their intimate knowledge of the springs, pitched in to help. Perhaps that was expecting too much of human nature, especially that of the WKPP leader, George Irvine. A muscular, combative, fitness-obsessed man, Irvine was another classic alpha who bitterly resented the Stone crew's ballyhooed invasion of "his" turf.

"There are a lot of people who don't like me, and I don't really care," Irvine told Norman, and one he especially didn't care about was Bill Stone. One of his posts on a widely read divers' Internet forum stated, "The bottom line is that in my opinion this guy [Stone] has no regard for human life, is a complete dillatante [*sic*], is a pretender, and has proven it for the nine years that I have been diving there while he has been yapping like a cocker spaniel." Stone refused to respond in kind.

Arriving at the springs, Stone's team found in front of their trailer a pile of trash bags with the note "You might need these for body bags." A bit later, divers found a dead catfish with a note stuffed in its mouth that read, "Triple this," a reference to Stone's prediction that his team, with its high-tech rebreathers, 3-D mapper, and fast electric scooters, might triple the existing mappage, which had taken WKPP ten years to accrue.

Then Stone's divers began finding that safety lines had been cut in a number of places.

They suspected that WKPP divers had done it, but Irvine called those charges "bull" and rejoined that Stone and his team were "a bunch of crybabies."

Nevertheless, the outsiders persevered. By March 1, when Stone's permit ran out, the team had made 3-D maps of every cave passage within the boundaries of the state park that served as their base of operations. That did leave a vast portion of the 450-square-mile cave system unmapped, but what the 3-D mappers lacked in quantity, they made up for in quality, proving that some of the handmade maps created by George Irvine and his teams were off by 300 feet in the first mile.

Stone doubtless awaited the publication of NGA's article with some trepidation. It did contain a few references (how could it not?) to Stone's relentless drive and brusque manner, but it gave legitimate credit to his inventions, his explorations—"an extraordinary accomplishment"—and his vision. As for WKPP, Norman let its leader, Irvine, come onstage and speak for himself, which brought no great credit to him or his organization. Stone held his tongue for the most part, which reflected well on him and his.

AS 2000 ARRIVED, STONE ONCE AGAIN began looking across the canyon at Cheve Cave. Nobody had come close to cracking Cheve's terminal sump, but at some point the choices in exploration shave cruelly fine: Which cave, mountain route, ocean deep, whatever, is 5 or 10 or 2 percent more likely to go? For several years, the odds had fallen in favor of Huautla. Now they were shifting back to Cheve, and so would Stone. He knew that "the Russians," as he called them, were going very deep in caves of their own, and he was determined to beat them.

He wasn't actually vying with Russians—that was a slip of the tongue caused by the Cold War tendency to call everything within the former U.S.S.R. "Russian." In fact, the opposition was Ukrainians, at work on a small, otherworldly plateau called the Arabika Massif in a small, strange eastern European region named Abkhazia. There, Ukrainian cavers had shattered the existing world record by a huge margin, descending 5,610 feet into a supercave called Krubera (KRU-bera) in the forbidding Caucasus Mountains, overlooking the Black Sea. The Ukrainians were just as skilled, determined, and courageous as the Mexican supercave teams. They made no secret of their belief that Krubera could go much deeper still. They typically mounted at least one expedition every year, and sometimes more, and would surely keep pushing themselves and their own supercave to the limits.

Thus the choice really had been made for him. Bill Stone would return not to Huautla and the Mother of All Sumps, but to Cheve. And this time he would not just be searching for the center of the earth. He would be racing to get there first.

TWENTY-SEVEN

LEGEND HAS IT THAT BEFORE EMBARKING on his 1914–16 South Pole expedition, the great British explorer Sir Ernest Shackleton placed this "advert" in London newspapers:

> Men wanted for hazardous journey. Low wages, bitter cold, long months of complete darkness. Safe return doubtful. Honour and recognition in event of success.

The story is probably apocryphal; historians have sought the actual ad unsuccessfully for decades. Regardless, more than five thousand men did apply for twenty-eight spots on Shackleton's *Endurance* team. That says something profound about the human spirit, or possibly about the unemployment level in pre–World War I Great Britain. Probably a bit of both.

One could imagine Bill Stone, preparing for his own 2003 expedition to Cheve, creating some variation of Shackleton's posting:

Participants wanted for journey to the center of the earth.
No wages, constant wet, cold, and darkness.
Weeks underground.
Safe return doubtful.
(Honor and recognition equally so.)

In reality, Stone's 2003 expedition advertisement was more sophisticated and upbeat. The United States Deep Caving Team (USDCT) published an impressive color brochure titled

Sistema Cheve 2003 Expedition
The All-Out Push to –2,000 Meters

The brochure was intended to attract both explorers and sponsors, and dangled this dramatic lure: "There is an enormous stake involved. . . . The expedition may, for all time, establish Cheve as the deepest cave on earth." How would this be done? By "using advanced life support equipment, radical climbing technology, and lightweight bivouac gear, the team will spend up to a month below –1,000 meters charting territory never before seen by humans."

It was an extraordinary prediction, given that in 2002 eight caves worldwide were deeper than Cheve. It was not even the deepest cave in North America, that honor still belonging to Huautla by a few feet. Regardless, Stone's expectations were both reasonable and supported by some persuasive evidence.

Jim Smith's 1990 dye-tracing experiment had established that water flowed without interruption from the mouth of Cheve all the way down to the Santo Domingo River, a vertical drop of more than 8,000 feet over a straight-line distance of 11.2 miles. What's more, the dye traveled all that way in just eight days, one of the faster transmissions cavers had observed. Such speed of transit suggested that the water was flowing not slowly through tight crevices but rapidly through big, open streamways. (Though *enough* tight crevices could also enable such flow.) Then there was the matter of alignment. After Cheve's first big vertical segment, the cave angled downward for several miles at a gradient of about one to ten, dropping one vertical foot for every ten horizontal feet traveled. If you projected a line dropping at that same gradient, it connected with the resurgence down in the canyon where the green dye had shown up. Finally, several as yet unconnected caves between Cheve and its resurgence lined up, more or less, between them. On the map, they looked like stitches in an incomplete seam. Stone believed that these were all part of one vast megacave, and that connecting them was the key to proving that Cheve was the world's deepest cave, once and for all.

ADVANCE ELEMENTS OF THE TEAM JUMPED off from Austin, Texas, on February 10, 2003. Establishing base camp was their first objective. It had been six years since a major expedition had lived in Llano Cheve. New growth had healed the brown scars left by earlier inhabitants, so a smooth field of green awaited. Expedition members and locals drove overloaded trucks as close as they could get to the llano, but a washout blocked them a mile away. There they parked, and for several days relays of cavers, antlike, made repeated carries from trucks to camp.

The Cheve 2003 team's core group—those whom Stone called his "rock stars"—arrived in mid-February. The Americans were Bart Hogan, from Maryland; Ohio's John Kerr; and Coloradoan Mike Frazier; Robbie Warke came from England, and Marcus Preissner from Germany. Caving is much closer to being a mainstream activity in Europe, and Poland contributed a contingent of rock stars with tongue-twisting names: Kasia Okuszko, Tomek Fiedorowicz, Kasia Biernacka, Pavo Skoworodko, and Marcin Gala.

The team's ages revealed something interesting. In 2003, Stone was fifty-one years old. Most of the cavers who had formed the 1994 Huautla expedition's core were about Stone's age then. All of them—Noel Sloan, Steve Porter, Kenny Broad, Jim Brown, Barbara am Ende, and others—were still caving. But none were doing the kind of extreme expeditionary caving that Huautla 1994 had involved, or that Cheve 2003 would. Stone, obviously, was still not only doing such work but pushing the envelope ever further, organizing and leading the expeditions that did it. The members of his team, however, were all younger. Hogan was forty-three, Kerr forty, Warke thirty-nine, Preissner thirty-four.

The divers formed a separate elite, a kind of Cheve Delta Force. "The point of the spear," Stone called them. The leaders were Britishers Rick Stanton, forty-two, and Jason Mallinson, forty, two of the world's best cave divers. Rich Hudson, another world-class Brit cave diver, and Stone himself were the backup dive team.

Another person would play an important role in 2003, a woman from

Alaska named Andrea Hunter. Just twenty-five (one wag noted that as Stone got older, his girlfriends got younger), she was coming on her first supercaving trip to learn the ropes and serve as a Sherpa. She had a master's degree in geology and was an expert mountaineer, diver, skier, and cyclist—a superb all-around outdoorswoman who, rather than being daunted by rough-and-tumble Alaska, positively flourished there. She was tall and tan, with shining tawny hair, freckles, sparkling green eyes, and a thousand-watt smile that rarely dimmed. Andi Hunter turned heads wherever she went, and that included the base camp of a Bill Stone supercave expedition.

When Hunter made her first descent all the way to Camp 3—a two-day trip—she found that Stone had cleared a double space for her in his camping area. He had a sleeping bag all laid out and hot soup waiting. Exhausted, she was touched by his thoughtfulness. Coincidentally, it was her twenty-sixth birthday, so the next morning, Stone gave her a present, an extra-large Snickers bar, and Polish cavers sang "Happy Birthday" to her in their language. She later remembered it as the best birthday she ever had, and it marked the beginning of her relationship with Bill Stone.

Andi Hunter was the latest woman in Stone's life after his marriage had ended. Barbara am Ende had been his first serious involvement. Their relationship survived the Huautla ordeal, but when Barbara's ardor for extreme caving cooled, so did Stone's—for her. After they split up, he became involved with a stunning brunette named Beverly Shade, an expert caver two decades his junior. Their relationship lasted about a year. Then along came Andi.

Before any of that happened at Camp 3, though, tremendous advance work was required. By March 9, the llano was a bustling village with colorful tent domes dotting the meadow. The "mess hall"—denoted by blue tarps hung from trees and the cliff face—housed worktables and big green Coleman stoves fueled by torpedo-shaped propane canisters. Containers of freeze-dried food had been sorted into small mountains.

John Kerr was a wiry, affable electrical engineer with a flat Ohio accent, a quick smile, and unbelievable stamina. He was also gutsy. Not long before, he had done Yosemite's legendary El Capitan—in reverse. Kerr had rappelled the entire length of the 3,000-foot face, then climbed back up using his caver's vertical rig. Climbers, after summiting, walk down a path on the great formation's back side. The most unnerving part, he said, was trying to pull up the tail of his rappel rope at the beginning, the better to control his descent. Since the 3,000 feet of it hanging below him weighed several hundred pounds, that was out of the question. The problem, with all that tension, was not slowing down but just getting started. He managed, by deft manipulation of his rappel rack.

Like everyone else, Kerr helped haul gear out of Cheve after the dive at the terminal sump was done. Of greater value, though, was his engineering expertise, which quickly made him the expedition's tech wizard. He fine-tuned the balky carburetor jets on the Honda generators for the 9,100-foot altitude and fired them up, lighting the camp and charging batteries. When new $1,200 custom-made light-emitting-diode (LED) headlamps intended to replace six hundred pounds of carbide wouldn't work, Kerr set up a field surgery tent and operated on them right there.

Laboring in base camp, Stone, Bart Hogan, and others stowed thousands of pounds of gear in cherry-red waterproof packs for later hauls into the cave, uncoiled miles of rope, and organized climbing hardware, scuba gear, power tools, and more. At the same time, a separate twelve-person team, led by Mexican cave veterans Matt Oliphant and Nancy Pistole, set up camp five miles north and about 3,300 feet lower in elevation than Cheve base. While the main team probed Cheve, this one, which was officially part of the overall expedition, would push a cave called Charco.

Charco was one of those stitches in the unfinished seam, lining up nicely between Cheve and its final resurgence down in the river. Linking Charco and Cheve would begin to create the megacave Stone believed extended all the way from Cheve's entrance down to the river. But—there was almost always a "but" in supercave exploration—Charco was hellish even by great cave standards. To reach its working end, a caver could spend three days crawling for miles through space like that under your kitchen table. Regardless, the potential link to Cheve made pushing Charco necessary.

A few cavers harbored a quirky kind of fondness for the awful place; they were called "diggers." The general population of supercave explorers breaks down into specialist subtribes: leaders and Sherpas, bolters and ropers and riggers, lead climbers and divers and diggers. In this rarefied context, "digger" is not a faintly pejorative label, like "laborer" or "trucker." "Digger," like "diver," conveys respect and gratitude for the performance of particularly dangerous, unpleasant, but necessary jobs.

Diggers are skilled specialists essential to supercave exploration. It is not unusual for a going cave passage to just stop. The most common blockages are breakdown and boulder chokes, but quite often they are not truly terminal. If the cavers can feel air movement coming through a wall of breakdown or boulders, it makes sense to try to find a way through, as Bill Farr had done in Cheve's earlier days. If no wormhole can be found, it makes sense to dig. But the diggers' greatest value is not laborious burrowing through the more obvious obstacles. Rather, it is their ability to find passages others cannot. Some people have inexplicable affinities with horses, engines, or children; the best diggers have that same kind of uncanny, instinctive feel for the subterranean earth, an ability to sense openings and going passages not remotely suspected by others.

John Kerr was one of the world's great diggers: "I'm happy in some little corner [5,000 feet deep] digging in the dirt with my titanium crowbar," he said in an interview. Diggers endure hours of confinement in squeezes so tight they can barely breathe and can move only one hand—places that would quickly drive most people slobbering, bug-eyed crazy. They dig like giant moles, using tools ranging from their fingernails to high-tech titanium implements to gasoline-powered hammer drills.

Riggers are another subtribe. While others established Cheve's base camp, the riggers began setting fixed ropes for the supply teams that would follow. Rigging is not just fixing rope for the big drops; a 30-foot fall can kill as surely as a 300-footer. In addition, many traverses—sections of wall that must be crossed horizontally, over yawning drops or roaring waterfalls—had to be rigged.

The little word "rig" does no justice to the sixteen straight days of work it took the team to fix two miles of rope, weighing a quarter ton, before the real Cheve exploration could even begin. The 2003 Cheve expedition brought eight hundred stainless steel anchors and bolts. It was a repeat of all the excruciating labor—the fixing of a double sequence of rebelay bolts—during the first rigging of Saknussemm's Well.

As they were rigging, the cavers were also "gardening," removing unstable rock for their own and the following teams' safety. As they descend, cavers use big hammers to whack away at any rock ledge or flake that looks dicey. Even when not gardening, cavers can be surprised. At one point, Bill Stone was hanging high on a wall in Cheve. Other cavers were milling around on the cave floor, hundreds of feet beneath him, their lights flickering like distant sparks.

Andi Hunter, Stone's girlfriend by then, stood on a ledge below, acting as his belayer. In other words, as Stone set bolts and climbed higher, he attached carabiners to the bolts and ran his climbing rope through them. Hunter, his belayer, held the other end of the rope. Just like an above-ground climber belaying a partner, if Stone took a fall, she would arrest it from her belaying stance. His drop would be limited to twice the distance between him and his last carabiner. If the rope between him and the belay was ten feet long, he would fall ten feet to the belay and ten feet below it.

Stone jammed his hammer into a small crack in the rock just above his head, wanting only to test its solidity. A tombstone-sized slab of cave wall peeled off and fell on him.

TWENTY-EIGHT

IT WAS LUCK OR THE GRACE of God, depending on your worldview, that prevented the rock slab, which weighed hundreds of pounds, from ripping Stone right off the rope. When it peeled off the wall, its trajectory plopped it perfectly in his lap. He clutched the tombstone and screamed "ROCK!" as a warning to the others. Then, feet braced against the cave wall, he shoved the slab out toward the middle of the pit in a desperate attempt to avoid hitting Hunter, right beneath him. She felt the *whoosh* as it passed very close, dropping 250 feet and exploding like a bomb on the cave floor.

Almost miraculously, no one was hurt or killed.

RIGHT BEHIND THE RIGGERS CAME THE Sherpas. Some were less experienced explorers earning their spurs doing grunt work, but all were expected to carry a load every trip down. Food, sleeping bags, pads, diving gear, batteries, stoves, fuel, tools, on and on. It was hard, dangerous, exhausting work.

One veteran of many load carries, David Kohuth, described it as "working like an animal. A mule. Just exhausted, always."

It was not unusual to see Sherpas rappelling down the 500-foot, vertiginous walls of Saknussemm's Well with two or even three red waterproof packs, each weighing up to forty pounds. There were those fourteen rebelays in Saknussemm's shaft, each requiring a stop to switch ropes. At times, many people were on the rope in that huge shaft, moving or halted at rebelays, their silver-blue LED lights glowing and floating like fireflies in mist.

John Kerr did yeoman Sherpa duty throughout the expedition, but his early trips, even with a relatively light load, were rough going. At one point, he was traveling down through the cave with a more experienced group, which, moving faster, inadvertently left him behind. (Given the constant noise of wind and water, and the intense concentration on route and rocks, and the enveloping darkness, that's not as hard to do as it might seem.) Alone, Kerr lost the route but didn't realize it. He arrived at what he thought was a wadable sump. Kerr strode right in. Suddenly the sump's bottom dropped away and, with a big pack on his back, he was swimming—*struggling*—to keep from going under. He made it to the other side, shaken. Later he learned that the proper route bypassed that sump entirely.

Having narrowly escaped drowning, Kerr was lost in one of the world's biggest supercaves. Eventually he heard another group passing through and joined them, but his initiation rites were just beginning. Not long after, alone once more (he hadn't been able to keep up with the second group, either), he found his way barred by a cylindrical, sheer-walled pit, called the Piston, with a river roaring through its bottom. The passage continued lower down, on the Piston's far side. To continue, Kerr had to traverse a section of vertical wall by clipping his carabiner to ropes running horizontally across it. When he got out to the midpoint of the traverse, the rope's sag put him just a few feet above the raging water.

At that critical point, his feet slipped off the wall and his heavy pack pulled him over backward, leaving him hanging upside down with his head a foot over the water. After a mighty struggle, Kerr was able to right himself and continue on. But if the traverse line had been a bit longer, putting Kerr's head into the river under all that pack weight . . .

BY MARCH 25, THREE CAMPS WERE in place. Camp 1 was an emergency bivouac 1,263 feet deep, not normally used. Camp 2, at 2,581 feet deep, was big and, as cave camps go, comfortable, but always windy. Camp 3 was located where it had been on earlier expeditions, a series of shelves high on one side of the cave, 4,078 feet deep and more than 5 miles from the main entrance. It took two days to travel from the surface to Camp 3. Regaining the surface was always a two- or three-day trip.

By this time, all of the camps were occupied. People streamed through them more or less continually on their way into and out of the cave. Standard operating practice here, as elsewhere, was hot bagging, described earlier. One woman caver recalled that the experience reminded her of some primitive mating ritual, lots of alpha males jockeying for female attention. With the ratio sometimes ten men to one woman, things could get very interesting.

In the same vein, plastic groundsheets did a good job of protecting sleeping bags from abrasion, but they had their drawbacks when it came to romancing on stone. The distinctive *crinkle-crinkle* sound they made, impossible to ignore, was Cheve's equivalent of creaking bedsprings. Some people were annoyed, but others were inspired. Before long, the entire campsite might be crinkling as though invaded by a flock of crickets. Though that kind of noise was unavoidable, participants did try to preserve some decorum by keeping the moaning and screaming down.

Cave cuisine was intended to maximize caloric intake; palatability was second. Cavers routinely burned six thousand to eight thousand calories a day. Stone, with precious little body fat to begin with, lost twenty-five pounds on some expeditions. Accordingly, they cooked up huge batches of freeze-dried and dehydrated food—dried meats, spaghetti, rice, potatoes—on their small stoves. To vary the glop routine, they snacked on nuts, jerky, chocolate, peanut butter, and candy bars.

With no way to dry soaked clothing or to clean mud off themselves (other than bathing in frigid underground rivers), the campers were always wet and always dirty. That made the risk of illness and infection greater, as did the fact that sunless environments depress the immune system. Those risks were increased even more by the cave's water, which flushed not only their own urine but also effluent from the surface, where the nearby towns' and villages' sewage disposal was primitive or nonexistent. All of this made tough work for the cavers' skin. After several days in-cave, it cracked and split, especially on the hands, giving microbes openings that, by their scale, they could drive trucks through.

Conserving light was just as important as saving weight, so unless they were working or hunting for the latrine, cavers turned their lights off and hung out in the dark, sleeping, tossing, turning, putting off the next latrine trip, or just . . . thinking.

Even just thinking could cause trouble in a supercave. Veteran Texas caver R. D. Milhollin made his first Cheve entry during the 2003 expedition. Reaching Camp 2, he and his three companions settled in to eat and sleep for a day or two. Wanting to conserve batteries, they kept their lights off except for essential tasks. After not very long, though, they confessed to one another that the cave was doing strange things to their minds. Milhollin was hearing phantom noises and seeing flashing lights. His girlfriend kept seeing skulls floating in the darkness.

They decamped promptly the next morning.

That experience, enough to disturb even the most stalwart souls, was actually mild compared to the *really* bad thing a big cave could do to one's mind, as Andi Hunter discovered.

TWENTY-NINE

HUNTER EXPERIENCED THE RAPTURE IN A neighboring cave called Palomitas. She went in with another woman, and after some time they came to the lip of a pit hundreds of feet deep. Hunter's partner rappelled away, leaving her alone on top. That was when The Rapture hit. Andi's heart started to race and she began to hyperventilate. Panic seized her. She could not remember how to use her vertical gear, or how to rappel, or even how to open her pack. She found it almost impossible not to stand up and bolt for the surface.

Hunter started to cry uncontrollably. She screamed down to her companion, who told her to sit tight and breathe, and it would pass. It did not. In fact, it got worse. She became absolutely convinced that she was going to die, then and there. Hunter had never experienced anything like The Rapture in her life, which had included frightening moments in other caves, as well as on mountains and dives.

In desperation, perhaps unconsciously, Hunter's mind turned to her mother. Aloud, she said, "Mom, if you're there, I really, really need you right now. *Please help me.*"

After a while, she began to calm down. She waited, and breathed, and kept thinking of her mother, and slowly her faculties and composure returned. Her poise regained, Hunter did the drop, joined her partner, and continued their trip. Had she been less experienced, The Rapture might well have killed her.

BEFORE LONG SHE WAS BACK TO work in Cheve. The purpose of everything, every bolt driven, wall climbed, pit rappelled, sump swum, pound hauled, and camp endured, was to put divers Stanton and Mallinson in the water. Describing Rick Stanton, an even-tempered English firefighter by profession, Stone bestowed the highest imaginable accolade: "cool and reserved, much like [Sheck] Exley." Mallinson, an equally skilled cave diver, bore some resemblance to the actor Mel Gibson. Every bit as proficient as his frequent partner Stanton, he was less a people person. Regardless, it was his work underwater that counted, not making nice in camp.

Mallinson, Stanton, and Rich Hudson, the third British diver, showed up in base camp on March 13. All three were surprised by how much the 9,100-foot altitude affected them, even just walking and doing light work. They were concerned that it might impair their diving or make them more susceptible to decompression sickness, but there was nothing they could do about it.

Other cavers had been rigging and stocking the camps long before their arrival, so the three readied their diving gear, then helped carry it to a staging area about 1,600 feet deep in the cave. They were not using Bill Stone's new and improved rebreather, the MK-V, which they thought too heavy and complicated, preferring to dive with their own smaller, simpler homemade rebreathers. In addition to simplicity, another advantage was that their units were "ventral mounts," meaning they were worn strapped to the chest rather than the back. That arrangement made squeezing and wriggling through tight spaces easier.

Stone himself would not dive on the Brits' stripped-down rebreathers. He was especially worried by the fact that they lacked redundancy and had no way to measure oxygen levels. To him, it was like driving a car with a good motor but no brakes. But he knew that Hudson and the others had logged scores of dives on their creations, and they were still among the living. If they wanted to go driving around without brakes, more power to them.

Eleven days after their arrival, Hudson, Mallinson, and Stanton arrived at Camp 3. Cheve's terminal sump, where they would begin their dives, was still several hours away, requiring passage through Wet Dreams, the Exclusion Tubes, and Nightmare Falls.

HUAUTLA 1994

February 1992: Left to right: Ian Rolland, Rob Parker, and Dr. Noel Sloan training with Cis-Lunar MK-II rebreathers in a hyperbaric chamber for the 1994 Huautla expedition's diving. Rolland and Parker, world-class cave divers both, later died in separate accidents, demonstrating the activity's extreme lethality. *(Photograph copyright © 2010 by Bill Stone)*

arch 1994: The United States eep Caving Team's 1994 uautla expedition in Oaxaca, exico. Standing, left to right: rbara am Ende, Paul Smith, m Morris, Wes Skiles, Kenny oad, Steve Porter, and James own. Kneeling: Noel Sloan eft) and Bill Stone. The only re team member not shown is n Rolland, who would also be e only one not to return. *hotograph copyright © 2010 Bill Stone)*

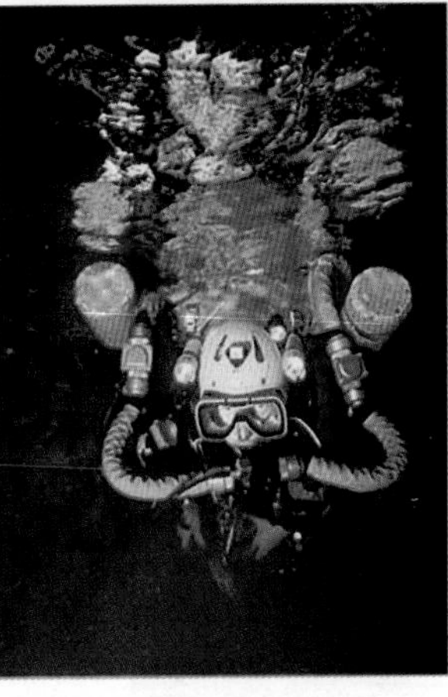

eft: April 1994: Camp 5, perched above Huautla's deadly San Agustín Sump, was about 4,300 vertical feet deep and miles from the cave entrance. An adjacent waterfall roared like a 747 engine. Screaming to make themselves heard, vers were constantly soaked and cold, and they slept on nylon hammocks strung from rock bolts above the portable atform. (Kenny Broad nearly died falling from one.) Ziploc bags were the latrines. *Right*: May 1, 1994: Barbara am nde about to dive the San Agustín Sump with Bill Stone. Ian Rolland drowned earlier using the same MK-IV breather. Am Ende and Stone's six-day penetration of Huautla, beyond all hope of rescue, ranked as one of the ntury's greatest, and least heralded, exploration achievements. *(Photographs copyright © 2010 by Bill Stone)*

CHEVE 2003

February 2003: Two miles of new rope, freshly washed, drying in Llano Cheve before rigging begins on the cave's ninety vertical drops, one more than 500 feet long. *(Photograph copyright © 2010 by Peter R. Penczer)*

February 2003: Bill Stone prepping vertical gear—rappel rack, helmet with specialized light he helped design, red harness, black caving boots, red waterproof pack. The expedition's blue kitchen tarp is in the background. The six-foot-four Stone started at 200 pounds and finished at 175. *(Photograph copyright © 2010 by Bill Stone)*

March 2003: Bart Hogan in Cheve's upper Entrance Chamber. The entire chamber could have contained three Boeing 757s. Behind Hogan, the floor dropped steeply to a large portal through which strong wind roared—the first sign of the cave's ultimate size. *(Photograph copyright © 2010 by Frank Abbato)*

March 2003: Bart Hogan on the natural altar in Cheve where ancient Cuicatec Indians sacrificed humans—including children—to their gods. *(Photograph copyright © 2010 by Frank Abbato)*

March 2003: Marcus Preissner rappels a section of Angel Falls, 1,100 feet deep in Cheve Cave. "Redirectionals" pull the rope away from plunging water. The immensity of supercaves and their absolute darkness make vast panoramic shots extremely difficult; the entire Angel Falls rappel was five times longer than the section shown here. *(Photograph copyright © 2010 by Ken Davis)*

March 2003: Marcus Preissner 3,000 feet deep in the vast Low Rider Parkway. Though impressive, the Parkway was not especially huge by Cheve standards. *(Photograph copyright © 2010 by Ken Davis)*

March 2003: Mariano Silva high-line traversing over the Swim Gym, below 3,000 feet. Later, John Kerr would almost drown here. *(Photograph copyright © 2010 by Gustavo Vela Turcott)*

March 2003: John Kerr squeezing out of Through the Looking Glass, the viselike 160-foot breakdown passage pioneered by Bill Farr. Just beyond yawned the A.S. Borehole, big enough to hold dozens of diesel locomotives. *(Photograph copyright © 2010 by Bill Stone)*

March 2003: Robbie Warke, 4,300 feet deep, about to rappel into mighty Nightmare Falls. Bill Stone, who climbed in Yosemite, remarked that exploring Cheve was like descending El Capitan, at night, in waterfalls, and then doing it all over again—upward—to get out. *(Photograph copyright © 2010 by Bill Stone)*

March 2003: Mariano Silva at Sump 1 in Cheve, 4,600 feet deep. John Schweyen first dove here in 1991 but was stopped by a constriction 330 feet in and 65 feet deep. The 2003 Cheve expedition's overarching goal was to "crack the sump." Further exploration might then prove Cheve the deepest cave on earth. *(Photograph copyright © 2010 by Gustavo Vela Turcott)*

March 26, 2003: British divers Jason Mallinson (left) and Rick Stanton make final equipment checks in Sump 1. Using homemade rebreathers, they forged a new route that passed the constriction which had frustrated all progress since 1991. Beyond Sump 1, they discovered almost a mile of narrow, steeply descending river canyon separated by huge lakes. *(Photograph copyright © 2010 by Bill Stone)*

On April 6 and 7, 2003, Mallinson and Stanton were joined by divers Rich Hudson and Bill Stone for a final twenty-seven-hour push. They passed Sump 2 but were eventually stopped by a ceiling collapse in an air-filled tunnel. The 2003 expedition cracked Sump 1, but did not establish Cheve as the Mount Everest of caves. *(Photograph copyright © 2010 by Bill Stone)*

CHEVE 2004

February 11, 2004: The 2004 Cheve expedition core team. Rear, left to right: Andi Hunter, Bill Stone, David Kohuth, and John Kerr. Front, left to right: Ryan Tietz, James Brown, Bill Mixon (whose Austin, Texas, home was the jump-off point), and Gregg Clemmer. *(Photograph copyright © 2010 by Bill Stone)*

March 7, 2004: The 2004 expedition produced more frustration than discovery. After others declared the Aguacate cave (possibly connected to Cheve) dead-ended, Andi Hunter fought through this squeeze and discovered a new route. It paid off in Aguacate with the discovery of more than a mile of largely horizontal tunnels leading toward the presumed junction zone with Cheve. This expedition encountered more tight spaces than giant vertical drops, which prevailed in 2003. Squeezes present their own unique hazards. Sometimes the only way to rescue irretrievably stuck cavers is to break bones. *(Photograph copyright © 2010 by Bill Stone)*

March 11, 2004: Andi Hunter spent five hours bolting this climb 150 feet above the Aguacate cave floor. Bolting was both dangerous and grueling. She used the twelve-pound drill to make three-inch holes in the wall. In the holes she secured stainless steel rock anchors, to which she then bolted steel hangers. From those she hung slings with carabiners and étriers (stirrups). Then she did it all again, gaining about 4 feet each time. (Before cordless drills, cavers did it all by hand.) *(Photograph copyright © 2010 by Bill Stone)*

March 21, 2004: Ryan Tietz dishwashing deep in the Aguacate cave. Light was the most precious resource. When not moving or working, explorers often turned lights off to save power. That preserved batteries, but could invite The Rapture. The 2004 expedition discovered 3 miles of new cave but did not establish the Cheve system as the world's deepest. *(Photograph copyright © 2010 by John Kerr)*

In 1976, Alexander Klimchouk, then nineteen, had already been exploring caves for eight years. He and a teenaged team established Uzbekistan's Kilsi Cave as the deepest in the Soviet Union, at 3,328 feet. *(Photograph copyright © 2010 by Marcus Taylor, courtesy of Alexander Klimchouk and the Call of the Abyss Project)*

1986: Alexander and his son Alexey at Kujbyshevskaja Cave on the Arabika Massif. In Eastern Europe, cave exploration is immensely popular. Children start caving as young as four. *(Photograph copyright © 2010 by Marcus Taylor, courtesy of Alexander Klimchouk and the Call of the Abyss Project)*

2004 CALL OF THE ABYSS KRUBERA EXPEDITION
August

August 2004: Base camp. Supercaving is like climbing Mount Everest in reverse. This expedition's fifty-six cavers from seven countries required five tons of supplies and needed four camps 2,297, 3,986, 4,593, and 5,381 feet deep. Two-week stays underground were not unusual. *(Photograph copyright © 2010 by Alexander Klimchouk and the Call of the Abyss Project)*

August 2004: Senior management. Klimchouk routinely delegated leadership. Here he confers with diver Gennadiy Samokhin (center) and supercave veteran Nikoley Solovyev. The three combined almost a century of extreme caving experience. *(Photograph copyright © 2010 by Alexander Klimchouk and the Call of the Abyss Project)*

August 2004: Rugged, self-effacing Yury Kasjan, then thirty-eight, rounded out the leadership. He first came to Krubera in 1989 and spearheaded many later expeditions. His occupation is "industrial mountaineering"—he uses supercaving's vertical techniques on skyscrapers and giant towers. *(Photograph copyright © 2010 by Alexander Klimchouk and the Call of the Abyss Project)*

August 2004: The entrance shaft drops 190 feet beneath Krubera's mouth. This perching caver weaves the rope through a bobbin rappel device while four others pay close attention. An incorrect rappel rack setup is called, not hyperbolically, "the death rig." *(Photograph copyright © 2010 by Alexander Klimchouk and the Call of the Abyss Project)*

August 2004: As in extreme mountaineering expeditions, 99 percent of supercave activity supports 1 percent at the sharp end—those exploring the deepest virgin territory. Here a "Sherpa" brings 160 pounds of supplies down a tight section of the 500-foot Big Cascade. *(Photograph copyright © 2010 by Alexander Klimchouk and the Call of the Abyss Project)*

August 2004: Supercaves' vast dimensions are one reason expeditions remain underground for weeks. Surveying is another. Every foot of new cave must be surveyed and recorded, as Julia Timoshevskaja is doing below 5,000 feet. *(Photograph copyright © 2010 by Alexander Klimchouk and the Call of the Abyss Project)*

In late August 2004, Gennadiy Samokhin returns from his history-making dive. In 32°F, zero-visibility water he pushed Krubera's limit to 6,037 feet, establishing it firmly as the deepest cave on earth. *(Photograph copyright © 2010 by Alexander Klimchouk and the Call of the Abyss Project)*

August 2004: Combat fatigue. Nikoley Solovyev, just up after fifteen days underground. (See the earlier "senior management" photo.) Cavers typically lost a pound or more a day in Krubera. *(Photograph copyright © 2010 by Alexander Klimchouk and the Call of the Abyss Project)*

Late August 2004: Gennadiy Samokhin, blinded by the light but ecstatic. Bravest of the brave, Samokhin was that great rarity among cave divers—one who was old *and* bold. *(Photograph copyright © 2010 by Alexander Klimchouk and the Call of the Abyss Project)*

October

October 2004: The smaller October expedition still required massive support: 2,200 feet deep, Igor Ischenko (left) and Emil Vash haul sixteen forty-pound packs down a 300-foot drop beyond the Lamprechtsofen Meander. *(Photograph copyright © 2010 by Ekaterina Medvedeva, courtesy of Alexander Klimchouk and the Call of the Abyss Project)*

October 2004: Ekaterina Medvedeva, then twenty-one, was born in Kiev and began caving at fourteen. The only woman on October's team, she lauded "the perfect atmosphere between expedition members." *(Photograph copyright © 2010 by Ekaterina Medvedeva, courtesy of Alexander Klimchouk and the Call of the Abyss Project)*

October 2004: Left to right, Igor Ischenko, Emil Vash, and Kyryl Gostev below 3,000 feet. Vash, then twenty-two, worried about doing well in Krubera. Exhausted and discouraged, he wrote at one low point, "I'm in the Dragon's asshole." Regardless, Vash performed admirably. He was also the expedition's most gifted chronicler. *(Photograph copyright © 2010 by Ekaterina Medvedeva, courtesy of Alexander Klimchouk and the Call of the Abyss Project)*

October 2004: Bernard Tourte passing through the nightmarish, 330-foot long Way to the Dream Meander. Discovered earlier by Dmitry Fedotov and Denis Kurta, it was ugly and tortuous, but it opened the way to the last great terrestrial discovery. *(Photograph copyright © 2010 by Sergio Garcia Dils, courtesy of Alexander Klimchouk and the Call of the Abyss Project)*

October 2004: Igor Ischenko perching over the entrance to a sump at 6,170 feet, close to the bottom of the world. Though he holds a flashlight, in 2004 Ischenko and most European cavers still used traditional carbide lamps as primary lights. *(Photograph copyright © 2010 by Ekaterina Medvedeva, courtesy of Alexander Klimchouk and the Call of the Abyss Project)*

October 18, 2004: Bottom of the world, 6,825 feet deep. Yury Kasjan (left) and Igor Ischenko, photographed by Ekaterina Medvedeva. Like other ultimate places—Everest's summit, the Challenger Deep, the North and South Poles—it was unimpressive to look at, as the discoverers' quotidian name suggested: Game Over. Nevertheless, the last great terrestrial discovery's significance was beyond measure. Henceforth, explorers would have to look off-planet for ultimate discoveries. *(Photograph copyright © 2010 by Ekaterina Medvedeva, courtesy of Alexander Klimchouk and the Call of the Abyss Project)*

October 18, 2004: The thrill of victory—sort of. Making the last great terrestrial discovery left even redoubtable Yury Kasjan too exhausted to smile. Indefatigable Ekaterina Medvedeva looked happier, but both knew that the worst was yet to come—getting out. *(Photograph copyright © 2010 by Alexander Klimchouk and the Call of the Abyss Project)*

THIRTY

AFTER SEVERAL MORE DAYS SPENT CARRYING gear down to the sump, Mallinson and Stanton were ready to dive. They would be the first to explore Cheve's Sump 1 since John Schweyen, in 1991, had penetrated about 330 feet and to a depth of about 75 feet, before the narrowing fissure passage stopped him. From the outset, Bill Stone had not intended to repeat Schweyen's route. He and Rick Stanton had studied the area's geology, and both felt that at some time in the past a fault had shoved the tunnel eastward. By starting off in the direction opposite from Schweyen's, they believed, divers could bypass the constrictions and the tunnel would eventually resume the desired direction.

Preparing the gear was critically important, but no more so than preparing the mind. There was no type of diving in which the risk of panic was greater. Some felt certain that this panic had killed the veteran caver Rolf Adams in Hole in the Wall Cave, and stories were legion of cave divers who had drowned with plenty of air in their tanks. Knowing this, Stanton in particular had evolved a protective ritual he performed before every dive.

He found a quiet place and visualized the route, using experience gained from his many previous dives to anticipate problems and prepare solutions for them. It was essential to avoid tension and "target fixation," as pilots call the tendency to focus on one thing to the exclusion of other important inputs. Stanton took himself through the dive's critical phases, working out every detail—tying off guidelines, doffing gear to push through tight squeezes, maintaining perfect buoyancy control. He also tried to anticipate things that might go wrong and to mentally rehearse solutions—cutting through line entanglements, using his gap reel if lost, switching to alternate regulators.

After donning his black dry suit, weight belt, and red helmet with three yellow and green lights attached, Stanton was first in the water. Sump 1 was a circular pool about 30 feet in diameter, with turquoise water and ridged stone walls the color of brass. Near its edge, the pool was about waist-deep. The water temperature was 64 degrees. There was no appreciable current. Standing there, Stanton strapped the rebreather to his chest. With Rich Hudson's help, he went through his lengthy predive checklist. He spit into his mask to prevent it from fogging, rinsed it, put on the mask and his bright yellow fins, and waited for Mallinson, who was soon in the water and ready. With final "okay" signals to Hudson, Stone, and others in the sump, they submerged and disappeared.

Stanton led, finding the route. Mallinson followed, running white safety line out from a black reel. The water was cold, but they were comfortable in their dry suits. The visibility was 6 to 8 feet. Their headlamps bored small tunnels of light into the dark water as they frog-kicked slowly forward. Before long, the tunnel turned almost 180 degrees and they were swimming in a direction that they'd figured should take them beyond the sump's previous terminus. Stone had been right. After about 80 feet they reached an airbell, which they passed through. They were 25 feet deep and 197 feet in when they swam over a hole in the tunnel's floor. It was tempting, but they opted to leave it for later. At 361 feet in they found, in the tunnel's right-hand wall, a hole big enough to swim through. Both divers immediately understood the significance of this "window." They slipped through the hole and surfaced 35 feet later in the main river passage continuation. They were greeted by the roar and crash of cascading water.

Stanton and Mallinson had just found a way past the terminal constrictions that had thwarted John Schweyen in 1991 and all progress for the years since. In so doing, they'd created the very real possibility that Cheve would go, and keep going, all the way down to its fabled resurgence at the bottom of Santo Domingo Canyon. And if it did that, Cheve, the long-shot candidate for world's deepest cave, would achieve that status by a very long shot indeed.

The two divers shucked out of their gear and started walking where no humans had ever gone before. Scooping booty, the cavers' dream. It was dicey terrain, a tight, smooth canyon that descended very steeply, carrying the main Cheve stream down in a tumult of spray and whitewater. Several hundred yards brought them to two 40-foot waterfalls, which they bravely downclimbed without ropes. Viewing these falls later, even Stone was impressed, admitting that he would never have attempted them without a rope. Given the Brits' audacity, the double waterfall's eventual name, Mad Man's Falls, was appropriate. After the second waterfall, they found the passage enlarging and leveling. Here they rested, snacked, and talked excitedly about what they had discovered and what lay ahead. They moved on, still in high spirits. But dark and labyrinthine caves, perhaps more than any other of earth's realms, will delight one second and destroy all hope the next. Before long they came to a second sump that stopped them cold.

Disappointed, the divers dragged back to Sump 1, estimating that they had traveled about 1,500 feet and descended perhaps 150. (Their total distance traveled and descent distance, revealed by later survey, would turn out to be twice that.) Those two slick and tricky waterfalls precluded hauling diving gear down to Sump 2, so they swam back through Sump 1 to their waiting companions. They had been gone six hours, and had spent ninety minutes underwater and the rest exploring on the surface.

After resting on the surface up in Llano Cheve, on April 5 Stanton and Mallinson dove through Sump 1, fixed ropes on Mad Man's Falls, then carried scuba tanks and weights to Sump 2. The next day, they dove Sump 1 again, followed by Bill Stone and Rich Hudson. Stone was no mean cave diver himself, but Stanton and Mallinson were in a class by themselves, so, diving with their own rebreathers, they remained the spearhead. Mallinson led through Sump 2 while Stanton spooled out the safety line just behind him. The dive took them through a passage shaped like a flattened oval about 16 feet high, 10 feet wide, and 40 feet deep. They swam 950 feet and surfaced in a boulder-walled pool with no obvious exit. Not willing to give up after having come so far, they stripped off their dive gear and spent four hours searching for a way on, without success. Checkmate. Again.

THIRTY-ONE

ALL FOUR DIVERS RETURNED SAFELY TO Camp 3 and then to Camp 2, where they rested overnight before continuing to the surface, having been underground this time for six days and at Llano Cheve for almost five weeks. Several days later, Stanton, Mallinson, and Hudson flew home. Everything that had been done, Stone's year of planning and organizing and fund-raising, the many team members coming from all over the world, all the work of all those people, had been done for the divers, who had made three dives and gone home. So it went in expeditionary caving.

While Stone and his divers had been exploring sumps, other cavers had been looking for openings in the cave walls above those sumps. A team of Poles led by Marcin Gala climbed and probed for ten days. On their last try they bolt-climbed 130 feet and discovered one tunnel that continued until they were over the midpoint of Sump 1, far below. There breakdown stymied them, but air blew briskly through it and into the tunnel. That air had to be coming from somewhere.

On April 14, Bill Stone, Robbie Warke, John Kerr, Marcus Preissner, and Bart Hogan began to derig Cheve, working from the bottom up, reversing all the hard work of the first weeks to "clean" the cave of everything they had brought in: ropes, camps, refuse, everything except bolts. In one last-ditch effort to bypass the sump, Stone, Kerr, and Preissner bolt-climbed a new route up a dome 230 feet above the cave floor. At first they thought they had found going passage, for here, as in the other tunnel, they felt moving air. But after a short distance, boulders stopped them yet again.

All that frustration might have been enough to sour some explorers on Cheve for good, but not Stone. "None of this stuff up there was known prior to 2003. There was some contention that this might actually bypass over the sump but we just didn't have enough time to push it," he remarked at the time.

Not on this trip, anyway.

AND WHAT OF THAT CONCURRENT EFFORT down in Charco, the attempt to slip in through Cheve's "back door"? Even for world-class veterans, it was a challenge. Stone and a caver named Mark Minton had discovered Charco's entrance in 1989, and in the ensuing fourteen years many expeditions had pushed its bottom down to about 4,400 feet deep and its overall length to about 2 miles. What made Charco awful, though, was that unlike most caves, it did not enlarge with depth. The inviting, spacious entrance soon constricted to a crawl space, if that. The team's expedition report noted that much of Charco was "descending ramps, tight-canyon passages, low crawlways, short rope pitches, and some even lower crawlways mostly filled with water . . . a cheese grater."

Before long, it was taking the Charco explorers twenty-four hours just to reach the cave's "business end," so they began camping at a site established in 2000, 1.5 miles from the entrance and about 2,200 feet deep. It was nasty even by cave-camp standards. There was room to sit up, but not stand. Cavers slept on small ledges no wider than a typical TV screen. One wrong roll would drop them 40 feet to the cave river below. The roof of those claustrophobic sleeping ledges was coated with a thick layer of gypsum crystals, which showered down every time the explorers brushed the ceiling. But they kept working and before long pushed down to 3,280 feet, requiring the creation of Camp 2, an even less hospitable site. The expedition members set bolts and hung from the walls in hammocks.

While divers were the spear point in Cheve, diggers ruled in Charco. It was not unusual for a digger to be wedged head-down in a vertical hole so tight that he could neither wear a helmet nor turn his head. To dig, he extended one arm down in front of his face, clawed and scraped material from the front of the hole, and ferried it backward, using his other hand to push the dirt out behind. Working this way, the digger resembled a swimmer doing the sidestroke.

Behind the digger came an assistant, who filled a bucket or pan and passed it back. And so on. Inverted in the hole, a digger was showered continually with dislodged dirt that worked its way into nose, mouth, ears, and eyes. Because carbon dioxide is heavier than air, the hole's atmosphere became increasingly foul, requiring breaks to avoid blackout. Charco diggers spent *many* long hours like this.

The object was not just to shove dirt around, of course, but to break through into a new passage, and it was the possibility of such a breakthrough that kept the diggers working through conditions that, even by caving standards, were ugly. By the expedition's end, they had extended the cave to 4,192 feet deep and about 4 miles long—an astonishing achievement, given the conditions under which they had labored for almost ten weeks. But they did not connect Charco to Cheve, so the discovery of one megasystem still eluded them and Bill Stone.

THE DIVERS HAD GONE HOME, AND now it was time for everyone else to go home, too. The cave explorers of the 2003 Cheve expedition left nothing on the field, as they say in athletics. Completing her first real supercave expedition, Andi Hunter found that the experience had been unlike any other in a life filled with arduous adventures. An experienced mountaineer, she'd learned that supercaving really *was* like climbing Mount Everest in reverse, with the worst—the ascent—coming last. Any number of times during her climb out of the cave, Hunter found herself hanging on to a rope over some huge abyss, so exhausted that she had no energy, no thoughts, nothing left at all except fatigue and pain. It occurred to her more than once that it would be so easy, and a lot less agonizing, just to let go, fall down into the blackness, and die a merciful death.

She and the others kept going. Writ large, that kind of endurance over ten weeks of work added another 4,133 feet to Cheve's length and extended its depth to 4,869 feet, making it the deepest cave in North America. At the very end, judging from descriptions, many cavers were so spent that getting out of the cave was a close-run thing. Hunter's team, for instance, outstayed their food supply, so for two days they had just two candy bars to feed their group of five. When they finally made it to Camp 2, they became subterranean dumpster divers, rifling through seven-year-old trash. They were overjoyed to find moldy hunks of cheese, which they quickly devoured, and filthy old tuna cans, which they unhesitatingly licked clean.

But for it all, they had not proved that Cheve was the deepest cave in the world. After all that gulag labor and Bill Stone's planning, organizing, sponsor seeking, fund-raising, and third-world diplomacy, Cheve remained only the ninth-deepest cave. Stone had made his first trip to Mexico almost thirty years earlier. He had accompanied or led something like fifty expeditions since, his ultimate goal still unachieved. For just about anyone else, it would have been quitting time.

Not Stone. Not even close. As his 2003 expedition whimpered to a close, he was already thinking about how to make 2004's end with a bang that would be heard around the world.

He could not have known just then that another legendary scientist and supercave explorer, the Ukrainian Alexander Klimchouk, was planning to do the very same thing.

PART TWO
KLIMCHOUK

There are many aspects, some of which lie beyond the scope of rational thinking, which experienced cavers evaluate.

—Alexander Klimchouk

THIRTY-TWO

AT 11:40 A.M., ON AUGUST 23, 2003, Alexander Kabanikhin, a young Russian speleologist from Archangel, descended into Krubera Cave, located high in the western Caucasus Mountains in the Republic of Georgia. Kabanikhin was ferrying supplies to others already at the cave's 500-Meter Camp, 1,640 feet deep. After a couple of hours, he arrived at the Big Cascade, a 500-foot-deep shaft with several rebelays.

At the first rebelay, halfway down, Kabanikhin switched his European-style mechanical descender, called a bobbin (comparable to the rappel racks used by most Americans), from one rope to the next, then leaned back to begin his rappel. The rope popped out of his descending device and he started to fall, his headlamp beam lashing the cave walls as he dropped. Before he could even scream he struck a rock ledge that smashed his mouth. In quick succession, other impacts compound-fractured his left leg, fractured his pelvis, and broke several vertebrae in his back.

Kabanikhin had made a mistake similar to the one that killed Chris Yeager in Cheve: failing to close and lock a safety gate on his descending device. Because his rope ran through two carabiners between his harness and the descender, Kabanikhin remained attached to the rope, but the carabiners and descender did nothing to slow his air rappel.

There is an old saying in climbing that the problem is not falling—it's the landing that gets you. Kabanikhin's plunge turned that on its head. The fall was actually worse than the landing because the rope, to which he remained attached, kept smashing him into the wall. Ultimately, though it caused grievous injuries, the rope did prevent his death. It did not hang completely free all the way to the bottom but, with some slack, led to the next rebelay, 115 feet farther down. Kabanikhin fell that distance and would have kept on falling another 100 feet, to his certain death, were it not for the rebelay anchor, which stopped him, not quite dead but almost. He hung there, unconscious at first, and then, soon after, awake and screaming.

Surviving that fall was a small miracle in itself, given the length and the brutal impacts. Two reasons he lived were his helmet, which kept his skull from being smashed like an eggshell, and the fact that he'd severed no major veins or arteries, though before long he might have thought bleeding to death a preferable fate. Pain slows time, and hanging there, badly damaged, Kabanikhin learned something about eternity.

Sergio García-Dils, a friendly, sandy-haired Spaniard, had arrived at the 500-Meter Camp earlier that day, bringing fuel for the team's gasoline-powered hammer drills and other supplies. A search-and-rescue expert who trained Spanish military units, García-Dils and others at the camp heard Kabanikhin screaming between crunching impacts with the wall. The cave acted like a vast echo chamber, amplifying the screams so much that, initially, no one was sure whether they were coming from above or below. At first, the cavers looked down into the vertical pitches beneath the 500-Meter Camp, but Bernard Tourte, a handsome, wiry, black-bearded French caver, soon realized that the screams were coming from above. Two other cavers climbed 100 feet to the rebelay bolt where Kabanikhin was hanging. A trail of bright red blood above, glistening in their lights, revealed the path of his fall. Kabanikhin himself was covered in blood, his face badly cut by its impact with the rocks. He was in agony from multiple fractures, but conscious and able to speak. As gently as they could (not very, given the circumstances), they lowered Kabanikhin to the 500-Meter Camp and made him as comfortable as possible.

Kabanikhin's situation might have been a lot worse. Without the rebelay anchor that halted his fall, he would have dropped that last 100 fatal feet. And although his plunge was calamitous, the 500-Meter Camp was still less than a full day's climb to the surface for healthy cavers. Had the accident happened down near Krubera's bottom, Kabanikhin might well have died before rescuers could have saved him.

Even so, Kabanikhin's chances for survival seemed slim. He was in horrible pain and in deepening shock. His pulse raced at 180 beats per minute, an indication of the trauma he had suffered. He had lost a quart of blood already, and his fellow cavers were having trouble stopping the bleeding from that compound fracture. To make things worse, hauling him to the surface would require a rescue litter, and the expedition did not have one. Even if they could obtain one, many passages between the surface and the 500-Meter Camp were too tight for a litter to pass through. Those passages would have to be drilled and hammered and blasted open before the litter could even be brought down. On top of everything else, only one member of the expedition had rescue training.

THIRTY-THREE

OTHER THAN DEPTH, CHEVE AND KRUBERA caves have very little in common. Krubera is situated at 43 degrees of latitude, some 1,800 miles farther north than Cheve. It is located in the contested region of Abkhazia in southeastern Republic of Georgia, north of the Black Sea. There, once upon a friendlier time, chain-smoking Russian men in Speedos and more sensibly attired women frolicked in salt water that, at 77 degrees Fahrenheit, was just brisk enough to counteract their copious infusions of vodka. From the shore of the Black Sea, the land rises dramatically to the Arabika Massif, a large mountain mass that is part of the western Caucasus Mountains; Krubera is within the massif.

To reach the cave in winter, modern explorers travel by helicopter. That's not as easy a trip as it sounds. In January 2005, a helicopter full of cavers and supplies crashed when it caught a rotor tip in the snow while trying to land in low visibility near Krubera's mouth. Everyone on board was injured, several people, including the pilot, seriously. Only sheer luck saved that expedition from multiple fatalities even before anyone set foot in the supercave. When the weather cleared, another helicopter evacuated the battered party, and that winter's effort was scrubbed.

In summer, the cavers ride old army trucks along roads that, for the last few miles, are little more than goat paths, to an elevation of about 6,500 feet on the Arabika, which is one of the largest limestone massifs in the western Caucasus, if not in the world. These are serious mountains, with frequent, powerful storms during the long winters and continual avalanche hazard. In summertime, though, the land surrounding Krubera is stunningly beautiful, like something out of a J.R.R. Tolkien novel.

The massif's elevations range from 5,800 feet to almost 8,000, and flying over it at 50,000 feet, you can see that it roughly resembles a four-leaf clover. The Arabika Massif is not huge—about 8 miles in a straight line from the outer edge of one clover leaf to the outer edge of its opposite. The clover here is white rather than green, because the massif and its mountains are composed of whitish and unusually hard limestone. The rock is high in silicon content, which gives it a sandpapery feel and a sparkly look under certain angles of light.

The massif might have been created by someone using a giant ice cream scoop to carve steep-walled, bowl-shaped valleys out of the mountainsides. That someone was Mother Nature, and her scoops were ancient glaciers, which left behind a fantasyland of emerald-green valleys and bone-white peaks and ridges, landforms so varied and chaotic they look like a storming ocean frozen in mid-tumult. The cave's immediate environs bear a name as beautiful and rolling as the land: the Ortobalagan Valley.

Krubera Cave's entrance, on a hill in the Upper Ortobalagan Valley, is a remarkably small hole in the ground, about the size and shape of a big-wave surfboard. Both Cheve and Krubera are true supercaves, thousands of feet deep and with many miles of surveyed passages, but there the similarities end. Cheve resembles a gigantic L, with an initial shaft dropping about 3,000 feet and an elongated foot descending for almost 2 more miles at a moderate slope of roughly 10 degrees. Krubera, on the other hand, is 90 percent vertical, pitch after pitch connected by short passages called meanders. While noncavers would be terrorized by the yawning pits, one of which is 500 feet deep, experienced explorers look forward to their thrilling rappels, if not to their grinding ascents. They uniformly despise, however, "the fucking meanders," as they are most frequently called.

Speleogenesis—the way caves like Krubera are created—is to blame for such conformation. Water plunging vertically has much greater erosive power than water flowing horizontally. Falling water can carve enormous vertical chambers, while water flowing laterally in the same cave will remove much less material. Thus meanders are excruciatingly tight, especially in their first few hundred feet, where outflow from the vertical chambers is slowest and carves out much less rock, year after year. Traversing Krubera's many meanders is like crawling under your car (while half submerged in flowing water) for thousands of yards.

The caves' body temperatures are also different. Cheve is cool but relatively mild, with air temperatures that range from 47 degrees at its entrance to the low 50s down deep. Water temperatures match the air temperatures. To eastern Europeans, that's balmy. Alexander Klimchouk said bluntly, "Mexican caves are *warm*." Krubera, on the other hand, is *freezing*, the kind of cold that, when you're working with water and wind, can quickly put you into third-stage, potentially fatal hypothermia. The average air temperature at the bottom of Krubera's first pit is 32 degrees in summer and zero or below in winter; it's 40 degrees at the cave's bottom. The inflowing water is never much more than 32 degrees; unprotected survival time in water that cold is fifteen minutes or less. Given the airflow though Krubera, the windchill factor remains below zero degrees Fahrenheit. Krubera cavers protect themselves with multiple layers of high-tech undergarments and waterproof dry suits, but dry suits are never completely that, so the cavers are always damp and always cold.

Both caves' explorers like to say that *their* cave is the more challenging. In reality, Cheve and Krubera both present unique difficulties. Cheve is longer, with more rivers and waterfalls. Krubera is steeper, tighter, and much colder. Very few cavers have been in both, but those who have describe them, in respectful tones, like this: very different, very difficult, and very dangerous.

Locals have known about Krubera's mouth for almost a century. Its entrance chamber was first descended and documented by Georgian speleologists in 1963. Over the next two decades, other scientists looked into Krubera, but none could get through an impassable constriction at 290 feet. Several expeditions visited the cave after that, but were unable to add anything new. The consensus formed that Krubera was undistinguished and had no potential for real depth.

Then along came a man named Klimchouk.

TODAY, ALEXANDER KLIMCHOUK, PH.D., is fifty-three years old and sports a large, perfectly trimmed, salt-and-pepper mustache. He looks like a blend of Charles Bronson and Walter Matthau: craggy and handsome like the former, avuncular and a bit mournful like the latter. If God had set out to design Bill Stone's diametric opposite, he could not have done a better job than Alexander Klimchouk. Stone grew up with a Brady Bunch family in a tree-lined suburb. Klimchouk was raised, more or less, in a Soviet industrial city by a twice-divorced mom struggling with infant triplets. Stone stands six feet, four inches tall and weighs 200 pounds. Klimchouk is five feet, eight inches tall and weighs 150 pounds. Stone is a classic type A—brusque, impatient, rushing. Klimchouk, or "Father Klim," as younger cavers sometimes call him, is mild-mannered, soft-spoken, polite to the point of courtliness, deliberate in thought and motion. Since his divorce, Stone has had relationships with one beautiful woman after another. Klimchouk has been married to the same woman, the hydrogeologist Natalia Yablokova, since 1975.

The two men are just as different in their exploration philosophies. Bill Stone unquestionably appreciates teamwork but values leadership more, preferring to be at the top of any team he accompanies. He has gone on relatively few expeditions that he did not lead. Klimchouk appreciates leadership but values teamwork more, firm in his conviction that only "a big society of united people" is capable of performing the multigenerational work supercaving requires. He routinely delegates leadership responsibilities to trusted younger cavers.

Klimchouk was born in Odessa, a Black Sea port 260 miles south of Ukraine's capital, Kiev, in August 1956—coincidentally, the same year and month in which French explorers discovered the world's first kilometer-deep cave in their country's Alps. Klimchouk's father died when he was four. His mother remarried, moved with her new husband to Kiev, and, when Klimchouk was six, gave birth to triplets, two daughters and a son.

Klimchouk's mother was understandably preoccupied with her three new babies. She did not ignore Klimchouk, but neither was she able to keep a close eye on him. When she subsequently divorced her second husband, the triplets' father, things became even more difficult. Kiev was no idyllic suburb. There were gray factories aplenty, cranking out fishing trawlers, machine tools, medical equipment, and motorcycles, among other industrial products. Kiev also had its share of gray concrete apartment buildings, gray weather, and gray apparatchiks.

But Kiev did have a bright and beautiful side. Tidy beds of red and blue flowers (Ukraine's national colors during the Soviet era) bordered sidewalks and streets, many of which were shaded by grand old chestnut trees. There was a Lenin Street, of course, but also the more splendid Shevchenko Avenue, which honored the national hero Taras Shevchenko, a nineteenth-century poet and painter who fought Russian tyranny. Good restaurants served chicken Kiev and other fancy edibles. Art shows and theaters enlivened life throughout the city, bathers played in the Dnieper River, and Olympic-caliber athletic events took place at the 100,000-seat Central Stadium.

The Soviet system, for all its faults, tried to do right by children. Two mainstays provided support for young people throughout the Soviet Union in general and Kiev in particular. One was school. The other, arguably more important, was a Soviet-wide organization called Young Pioneers, a blend of the Boy Scouts and Girl Scouts, the YMCA, and the Young Republicans (or Democrats). The state-sponsored Young Pioneers offered more than American youth groups ever had. In the 1970s, for example, a Young Pioneer camp called Orlyonok hosted seventeen thousand youngsters each year. The camp had sixty buildings, a movie theater, outdoor and heated indoor swimming pools, a secondary school, a medical building, an air and space museum, a passenger ship, forty-five sailboats, motorboats, and two hundred different activity offerings.

Every city, Kiev included, had Young Pioneer Palaces, centers for recreation, athletics, education, and, of course, some good old-fashioned Communist indoctrination. The first such centers were created in Moscow in the 1920s. These actually were former palaces, from which wealthy and politically incorrect owners had been evicted, not infrequently with holes in the backs of their heads. The government also built new Young Pioneer Palaces. By 1970 more than three thousand of them functioned throughout the Soviet Union. Kiev's, completed in 1965, was one of the largest, on a par with Moscow's. The Kiev facility was a sprawling, modern four-story building with vast expanses of glass and bright, spacious interiors. Grassy pavilions with chestnut trees surrounded the campus. There was even a silver-domed observatory on the roof.

The palaces complemented the Soviet schools, which stressed academics and indoctrination almost exclusively. Palace membership was voluntary, and children could choose from age-classified "hobby groups" in sports, creative and artistic pursuits, technical instruction, politics, and various outdoor activities. No tuition was charged, and the quality of instruction was excellent. One common, and classically Communistic, responsibility linked everything: *teach what you have been taught.* That theme would prove especially beneficial for cave exploration in eastern Europe, where leaders inculcated their teams with the idea that unlocking the secrets of supercaves was a lifetime commitment whose success depended on passing skills and knowledge from one generation to the next.

In the end, though, Young Pioneering was more play than politburo. Part of the goal was to produce well-rounded youths. Another was to identify unusual potential in children and develop it for nationalistic purposes like the Olympics and scientific discoveries. Alexander Klimchouk displayed such potential.

For as long as he could remember, Bill Stone had been fascinated by science and, in his youth, by chemistry in particular. Klimchouk, almost from the time he learned to read, loved geology. With extraordinary initiative, at age eleven he embarked on a self-directed search for knowledge in the Kiev Young Pioneer Palace.

Klimchouk passed through the imposing front doors and walked hall after echoing hall, exploring each floor in turn, finally coming to an open door on the other side of which sat an old professorial type.

"I'm looking for the geologists," Klimchouk said.

"Come on in," the man replied with a smile and a wave. "Here we do speleology, but it contains also geology."

Seeing one image in a high school slide show changed the course of Bill Stone's life. Passing through that office door changed Klimchouk's.

THIRTY-FOUR

FORTUNE FAVORS THE BOLD, even if they are just eleven years old. As it turned out, the amiable geologist was none other than Valery Rogozhnikov, himself just twenty-two at the time (when you're eleven, twenty-two can seem very old indeed), the founder of organized caving in Kiev. Caving soon took center stage in Klimchouk's life. Finding himself part of what in the United States would be called a "grotto," or serious speleological club, in Kiev, he devoted every free moment to planning, organizing, and going on expeditions. His passion for caving led to repeated absences from school—some lasting weeks. Despite that, Klimchouk, an exceptionally intelligent youth, did well academically. He graduated from high school on schedule and took university entrance exams. Well, *a* university entrance exam. There were five required. He took one but blew off the other four to join friends leaving for some distant caving in Uzbekistan in the summer of 1972.

The main group went by train—four long days of travel. Klimchouk started two weeks later and traveled alone by hitchhiking. As in America,

1972 was a different time in the Soviet Union; hitchhiking was not only safe but easy. Most drivers, even without being hailed, would stop and offer a ride to a teenager trudging along the roadside. So after a transcontinental thumb trip, Klimchouk joined his friends in the mountains and worked with them for two weeks, until they left. Teams working here occupied a high, remote, windswept plateau, shaggy green and as flat as dead-calm water. They lived in a multicolored collection of A-frame tents that lined up as perfectly as any ever erected by a Russian army platoon. The cave mouths were a half mile distant, higher up on the ridges that rimmed the plateau on all sides.

Klimchouk stayed in the mountains, waiting for another group, which eventually arrived. He led this second group for another month of exploration. They discovered and explored to 864 feet a cave they named Kilsi. Accompanying three more expeditions over the next four years, Klimchouk worked with teams that established Kilsi, at 3,328 feet, as the deepest cave in the Soviet Union, making it the first kilometer-deep cave in the country and the fourth-deepest in the world. Not bad for teenagers using the most primitive equipment, including clumsy cable ladders, and developing new techniques as they went along.

Kilsi would eventually drop far down the list of the world's deepest caves, but Klimchouk's experience there was crucial to his later work. Kilsi proved that he and his teams had what it took to make world-class discoveries underground. The Kilsi experience also demonstrated something else, just as important: supercave exploration would be a long effort requiring unprecedented determination, endurance, and persistence. That might have put off some explorers and scientists, but not Klimchouk. Rather, it produced a vision of something that could become the fruitful and exciting work of a lifetime—generations, even. A *dream*, in other words: discovering the deepest cave on earth. Here was forged a commitment like the one Bill Stone had experienced, a dedication so powerful that it would determine the course of Alexander Klimchouk's life and affect the lives of many others.

A very deep cave was not the only discovery Klimchouk made on the 1972 Kilsi expedition. Another team member was a lively, lighthearted, red-haired young woman named Natalia Yablokova, who had actually started caving a year before Klimchouk. Natalia was a superb, very active caver as well as the perfect whimsical counterbalance to Klimchouk's own gravitas. They fell in love and married in 1975. Their first child, Oleg, born in 1977, got an earlier start caving than any of them. Natalia was pregnant with him when, as part of another expedition in 1976, she descended 1,200 feet in Kilsi Cave. After Oleg's birth, Klimchouk completed his two years of obligatory military service, from 1977 to 1979. Their second son, Alexey, was born in 1982.

In 1979, Klimchouk established the Institute of Geological Sciences at the National Academy of Sciences. Before long, he was directing a dozen scientists and technicians. One of those scientists under his supervision was Natalia, who joined the Karst and Speleology Department in 1981 and worked there for the next decade. (While cave exploration, sadly, destroyed Bill Stone's marriage, it only strengthened the bonds of Klimchouk's.) Overseeing the work of such a large group of scientists would normally require a doctorate. Klimchouk completed his "on the fly," as it were, his progress interrupted by frequent expeditions; in 1998 he finally received his doctorate in hydrogeology.

By this time, Klimchouk's involvement with caves had two sides, closely related. One was with the voluntary caving movement—clubs, societies, associations. In 1975, he became the chairman of the Kiev Caving Commission, the body that coordinated the activities of several caving groups in the city. In 1984, he founded the Kiev Speleological Club, which united most of the caving groups into a single organization with about one hundred members.

The other line was in-the-ground sophisticated scientific inquiry involving karst hydrogeology, focusing on the origin of caves and on the subterranean water sources called aquifers. The study of aquifers is a major branch of science unto itself, which stands to reason since they are critically important to two of the most basic human needs: food and water. Virtually all agriculture, from kitchen gardens to massive megafarms, requires irrigation with water drawn from aquifers. Fully as much as their vegetables, humans in settlements from tiny African villages to megalopolises like Tokyo depend on aquifers for their water supplies—ergo, for their very existence. Other than its air, it is hard to imagine anything disrupting a modern city more quickly and completely than a break in its water supply. There are ways around living without electricity and gasoline and even easily available food, but not without water. By all accounts, terminal dehydration is one of the most excruciating ways to die, and it is not too great a stretch to visualize thirst-crazed city dwellers drinking their neighbors' blood, as shipwreck survivors in lifeboats and other victims of extreme dehydration have done.

The topic of science sheds some light on Klimchouk's and Bill Stone's different approaches to supercave exploration. Both men indisputably enjoy the thrills and adventure of supercave exploration, though the need to raise funds from buttoned-up corporations has sometimes led them to downplay what might be perceived as the less serious aspect of what they do. (Those who don't thrive on life-and-death excitement tend not to last very long as cave explorers, in any case.) But both men also have loftier goals and different means of trying to achieve them. It can be said that for Stone, science became a means to an end: exploring. His Ph.D. in structural engineering was an ideal scientific preparation for inventing the sophisticated rebreather, which in turn made a whole new epoch of extreme cave exploration possible. It would be untrue to say that Stone has no interest in scientific discoveries. He most assuredly does. But it is also true that he sees himself as the forerunner—"pioneer" is the term he likes—who opens new realms where others can follow to push their own frontiers in biology, chemistry, geology, psychology, paleontology, and more. Of course, the thrill of adventure is inherent in pioneering, and Stone was by no means immune to its allure.

For Klimchouk, it is the other way around: science is the end and caves are the laboratories in which he does it. "I was always on the scientific side," he likes to say. The two explorers' publications reflect these differences. Stone's caving expeditions, for better and sometimes worse, have gained their widest exposure in mainstream publications such as *Outside*, *National Geographic*, and *National Geographic Adventure* magazines. Similarly, his 2002 book, *Beyond the Deep*, coauthored with Barbara am Ende and Monte Paulsen, was intended for the general reader. He has written more than 100 caving-related articles, but they have appeared in publications such as the AMCS *Activities Newsletter* that are more for enthusiasts than scientists or academicians. His articles for professional, "pure" scientific journals and conference proceedings (more than 140 at last count) have been about engineering topics rather than caving.

Klimchouk, in contrast, writes largely for professional scientific journals, having realized early on that as a scientist who caves he "was able to produce discoveries that are globally important." He published his first scientific article when he was fourteen (in a now-defunct Soviet geology journal) and has authored scores since. They have titles like "Unconfined Versus Confined Speleogenetic Settings: Variations of Solution Porosity and Permeability" and are, to say the least, highly technical. The one publication where the two explorers' paths have crossed is *National Geographic*, which has published articles about expeditions led by both men; but otherwise they write for different audiences.

Another difference has to do with the explorers' respective approaches to organization. Bill Stone, with an almost pathological dislike of bureaucracy, prefers to operate on his own terms, unfettered by rules and regulations. He tends to steer clear of organizations other than companies or government agencies that can provide funding.

Alexander Klimchouk, having sprouted, as it were, from the fertile ground of such Soviet organizations as the Young Pioneer Palaces, has thrived on affiliation. Seeing the need for better organization and stronger leadership, Klimchouk chaired the Kiev Caving Commission and then created the Kiev Speleological Club, joining together smaller, disparate groups to foster larger, more ambitious, better-organized explorations. Later, he would escalate from a city-level organization to a national one, creating the Ukrainian Speleological Association.

Having cited so many differences between Stone and Klimchouk, I would be negligent not to point out a very salient similarity: both men are competitors, and from the beginning both understood that, science aside, the search for the earth's deepest cave was indeed a grand competition—a race. Stone described it as such in so many words. Klimchouk, for his part, used almost the same phrasing: "Cavers compete for the discoveries." Perhaps feeling a bit abashed after that declaration, the Ukrainian added, "You know it's part of human nature."

In 1980, speleological consensus was that the Arabika Massif had, to use the mining term, played out. Lots of promise, no payoff. Klimchouk thought otherwise but needed more than intuition to marshal the resources for further exploration there. In 1984 and 1985, he poured fluorescein dye into several caves, including Krubera, high on the Arabika. Traces of that dye later flowed out of springs on the shore of the Black Sea far below. More traces tinged the water 400 feet beneath the surface of the Black Sea, miles offshore. Klimchouk's dye traces proved that this was the world's deepest karst-based hydrological system.

He directed or participated in most of the work being done by the Karst and Speleology Department at the Institute of Geological Sciences and the Ukrainian Institute of Speleology and Karstology. He and his teams conducted research in the field, especially western Ukraine, but also in Russia, Armenia, and central Asia. These were among his happiest years, perfectly blending exploration and legitimate science.

The Soviet Union's 1991 dissolution ended all that. In its aftermath, Ukraine suffered terrible unemployment, inflation, and growing crime. In a country struggling just to survive, there were few resources for "unpractical" things like cave exploration. With everything from academia to zoos in turmoil and without funds for salaries or research, Klimchouk was forced to dissolve his department. Suddenly something like two dozen senior researchers found themselves cut adrift.

One of these was Klimchouk's wife, Natalia. She, however, had a fallback position and, not surprisingly, it involved caves. For some years she had been deeply involved with children's caving groups. Since 1986 she had led the children's caving group of the Kiev Speleological Club, which she continued to do. In 1992, she took charge of the Children and Caving Commission of the Ukrainian Speleological Association. She would go on to serve as copresident of the International Union of Speleology's Children and Caving Working Group. Just as Cub Scouts and Boy Scouts in the United States have den mothers, for several generations of cavers in Kiev and Ukraine, Natalia Klimchouk was their "caving mom."

If he could not save the academic and scientific organizations, Klimchouk strove to preserve the speleological movement in general. He founded the Ukrainian Speleological Association (Ukr.S.A.) in 1991, a national caving organization that united most of the groups and clubs in the country. The Ukr.S.A. enabled speleological activity to survive through the crises of the 1990s. It also created a major source of training for cave explorers and maintained motivation for the deepest cave search.

Within a few years, the Ukr.S.A. had over five hundred members and was a going concern, publishing magazines and newsletters and organizing regular seminars, conventions, field training for vertical and rescue technique, and multiclub expeditions. With the number of active senior cavers decreasing, it was critical to pass on traditions, skills, and an understanding of the importance of finding the deepest cave. The Ukr.S.A. delivered this inheritance. Because the Ukr.S.A. was (and still is) the only multinational speleological organization created in the post-Soviet countries, many cavers from these countries, especially from Russia, joined up.

Around the turn of the millennium, Ukraine's government became (relatively) stable, its economy a few steps back from the brink. It seemed a good time for Klimchouk to revive his scientific group from the ashes of communism's collapse. Here he devised a strategy as ingenious in its own way as Bill Stone's caving-for-course-credit pitch to Rensselaer Polytechnic Institute.

Reasoning correctly that a big pot would collect more gold than lots of little ones, Klimchouk united all of Ukraine's karst and cave scientists into one entity, the Karst Institute. It was an impressive group in both size—more than thirty respected scientists working at different institutions and universities—and prestige. Then he approached not one but two government agencies, the Ministry of Science and the National Academy of Sciences, which were competing for national primacy and international recognition. Klimchouk pitched both, then stood back and let the miracle of capitalist-style competition work its magic. Before long, both agencies were striving to be the Karst Institute's most liberal patron, invigorating Klimchouk's creation with funds and powerful political support.

THIRTY-FIVE

ALEXANDER KLIMCHOUK WOULD MAKE THE ARABIKA Massif world-famous, but he was not the first speleologist to set foot there. That honor went to an extraordinary Frenchman named Édouard Martel. A lawyer–turned–cave explorer who is generally acknowledged to be "the father of speleology," Martel came in 1902. He was traveling on a press junket arranged exclusively for him by the Russian government, anxious to stimulate Black Sea tourism. Martel obligingly published an account of his travels, titled to make readers associate the region with one of Europe's legendary vacation spots: *La Côte d'Azur Russe* (The Russian Azure Coast). Chapter 16, "L'Arabika Massif," described his visit to the Ortobalagan Valley, where he explored a cave, known today as Martel's Cave, whose mouth is only a few hundred yards above Krubera's.

Close on Martel's heels in the Arabika Massif came the Russian scientist Alexander Kruber. If Martel's visit had more to do with boosting tourism, Kruber's interest was purely scientific. During 1909–10, Kruber performed field studies and issued a number of publications about his findings. For these and other geological works he is regarded as the founder of karst science in Russia.

The Russian Revolution, two world wars, global economic crises, and countless regional conflicts diverted eastern European attention from cave exploration for five decades. Finally, in the 1960s, Georgian scientists began looking anew at Arabika's possibilities. Despite primitive equipment and vertical techniques, these early explorers managed to descend about 700 feet in several Arabikan caves, leading them to believe that the massif warranted further investigation. Ironically, it was one of the shallower descents that would ultimately validate their expectations. This was an open-mouthed, 200-foot shaft penetrated in 1960 by a scientist named Leonid Maruashvili. Perhaps sensing the pit's potential, Maruashvili named it Krubera, after Alexander Kruber.

Tantalized by the Arabika Massif's potential, over the next two decades a number of expeditions came, saw, and kept right on going. Their explorations appeared to contradict rather than confirm the plateau's promise, revealing no caves deeper than about 780 feet. By the late 1970s, virtually all speleologists had turned their attention elsewhere.

Virtually, but not *all* speleologists. Enter, in 1980, Alexander Klimchouk, now twenty-four and the leader of the same Kiev Speleological Club that had provided his first taste of caving thirteen years earlier. Klimchouk was a rising star in the world of cave science and already committed to finding the deepest cave on earth. Rather than join other clubs working elsewhere, Klimchouk chose to focus on Arabika.

It's hard to exaggerate the importance of that decision, and hard, as well, to fully explain it. Certain people seem blessed with a genetic affinity for particular phenomena, whose miracles they reveal for the rest of us. Edison understood electricity with every vibrating cell in his body. Amundsen, an icy man himself, was as at home in the polar scapes as white bears and black orcas. Alexander Klimchouk, who began going underground when he was eleven and (as noted earlier) published his first scientific paper at fourteen, had such affinity with the subterranean world.

That intuitive faith—call it a sixth sense, for lack of a better term—kept pulling him back after the entire speleological community had abandoned Arabika and moved on. Sounding as much like a mystic as a scientist, he would ultimately attribute the attraction to "*other, sometimes mysterious, feelings*." (Emphasis added.)

There were also concrete reasons, of course. As a karst scientist, Klimchouk saw one great advantage in Arabika: thick layers of limestone stepping continuously from those high mountains all the way down to the Black Sea. Those layers held out the possibility that Arabika's caves might also go all the way down to the Black Sea, 8,000 feet lower. This flew in the face of the conventional wisdom, which held that the central caves of Arabika could *not* be hydrologically connected with the shore because of other geological barriers.

Finally, Klimchouk thought he knew the reason why previous cave explorers in Arabika had come away empty-handed. Traditionally, their investigations had been, both literally and figuratively, quick and dirty. Patient by nature, made thorough and meticulous by training, Klimchouk was suspicious of that superficial approach, which he called "quicksearch."

Quicksearch started, literally, at the top. Speleologists of the day tended to look for cave entrances in big sinkholes because (as Carol Vesely and Bill Farr knew) sinkholes often signal the presence of caves beneath. However, in the high Caucasus massifs, glaciers had dramatically altered the original karst landscape, which had indeed been riddled with sinkholes. Glacial scouring removed the uppermost limestone, including most of those sinks, and filled many that remained with glacier-transported debris. Understanding this, Klimchouk knew that more attention should be paid to small entrances and fissures, located not within sinkholes but . . . *everywhere*. This obviously required not quicksearch tactics but efforts both exhaustive and systematic.

The quicksearch philosophy reigned not only above ground but below it, as well. After finding an opening, speleologists then made their way into a cave until something—a boulder choke, a squeeze, an underground river, whatever—stopped them. Rather than beat heads and hammers against such obstacles, they just moved to new shafts and passages. The underlying assumption was that there were hundreds, maybe thousands, of holes in the ground around here. If one didn't work, it was easier to pick up and move on to a new one than to slave away like hard-rock miners. Klimchouk and his teams would not turn away from obstructions. Instead, they would adhere to a "no dead ends" philosophy.

That thinking paid off as early as 1980, when Klimchouk demonstrated that in Arabika, sinkholes were not the only way down and, as well, that size really did *not* matter, at least on the surface. That year, he became intrigued by a cave mouth, not much bigger than your toilet seat, lurking beneath a crumbly limestone ledge. Other explorers had seen it and kept right on going. Klimchouk, however, thought it was worth a closer look, and how right he was. That tight mouth proved to be the beginning of what is now known as Kujbyshevskaja (KOO-bye-chef-Sky-ya) Cave, almost 4,000 feet deep.

Klimchouk saw that it was important to abandon quicksearch not only underground but in relation to time as well. He understood that of all the earth's geographic and natural phenomena—mountains, oceans, rivers, the atmosphere—caves were the least likely to yield ultimate secrets to casual suitors. They would demand more grueling, deadly, and unrewarding persistence than any other terrestrial feature, persistence that might well require not just years but *generations* of exploration.

Inspiration, of course, was only one of discovery's parents; perspiration was the other. Klimchouk and his teams of cavers put their sweat, and sometimes their blood, where their theories were. In one instance, it took them three years, from 1983 to 1986, to clear out just one boulder choke 2,200 feet down in Kujbyshevskaja Cave. They called this obstruction Ugrjum-Zaval, a name as ugly as the work required to clear it. Ugrjum was a very big boulder choke, a 300-foot vertical shaft 10 to 15 feet in diameter completely filled with rocks and boulders and pouring with 32-degree glacial meltwater. On the surface, digging of that magnitude would be done with big Caterpillar power excavators and clamshell cranes. Down deep, humans did the work with hand tools, ropes, pulleys, and endless determination.

IN THE ORTOBALAGAN VALLEY, KIEV CAVERS made dramatic progress. In Kujbyshevskaja Cave, they pushed below 3,500 feet in 1986 through a series of boulder chokes—Ugrjum-Zaval was just one—previously labeled "hopeless." In the early 1980s, Klimchouk teams' discoveries in the Ortobalagan Valley attracted other caving organizations from the former Soviet Union. Coordinated by Klimchouk, the whole Arabika Massif was divided into discrete search areas, and all expeditions adopted his systematic, "no dead ends" approach. By the end of the 1980s, some thirty-six Arabikan caves deeper than 300 feet had been explored, including seven caves deeper than 1,750 feet and three caves deeper than 3,300 feet. In addition, the dye-tracing experiments Klimchouk conducted in 1984 and 1985 proved that there were indeed hydrological links with resurgences on the Black Sea shore, thus confirming his earlier suspicions and revealing the world's foremost potential for supercaves.

In Krubera Cave, Klimchouk's teams almost unlocked the whole Arabika Massif system. A vertical section in Krubera called P43 (because it was a pitch of 43 meters, or 141 feet) began about 700 feet deep and dropped another 140 feet. Between 722 feet and 820 feet cavers identified, but did not penetrate, two openings, called "windows," in the cave wall. This was not a violation of the "no dead ends" system. As long as the cave kept going down, it made sense to follow its path of least resistance. But when they could go no deeper in Krubera, Klimchouk and his cavers decided to see where the two windows might lead. Opportunity was beckoning beyond those two dark portals. It was time to go through them.

And so they would have done, were it not for the war.

THIRTY-SIX

IN 1990, THE SOVIET UNION BROKE up into fifteen separate republics, some of which kept disintegrating into even smaller pieces. The Republic of Georgia was one that aftershocks continued to crack, enthusiastically aided by Russia.

In 1992, the province of Abkhazia, abetted more or less openly by Russia, went to war to gain its independence from the Georgian republic. Their "country" was landlocked within Georgia, but ethnic Abkhazians felt more affinity with Russia and declared themselves a separate, sovereign nation. Georgia sought to preserve its union. It was the latest eruption of an ethnic conflict that had been going on since the Middle Ages, and it retained much of that epoch's brutal flavor. The warring factions killed and tortured each other's soldiers, old people, women, children, babies, dogs, farm animals—anything that moved, really, for two years. Both sides perpetrated atrocities with gruesome ingenuity and lack of discrimination about targets. That made

Abkhazia one of the best places on earth for outsiders, including even Ukrainians, not to be.

Among the war's many casualties was cave exploration on the Arabika Massif. The conflict "officially" ended in 1994, the same year that Bill Stone's Huautla expedition produced such tragedy and triumph in Mexico, but the killing and atrocities never entirely died out, flaring up periodically here and there like small, deadly fires in a burned-over forest.

Ukrainian cavers finally returned to the Arabika Massif in August 1999, although conditions in Abkhazia remained unstable. This was a Ukrainian Speleological Association expedition, organized and led by Yury Kasjan, a rugged and self-effacing Ukrainian, then thirty-eight, who was among the world's top cavers and who would thereafter play a vital role in the search for the deepest cave.

Unlike Bill Stone, who led most expeditions he accompanied, Klimchouk routinely delegated leadership to carefully chosen, younger veterans, leaving himself the role of strategist and doyen. He had several reasons for doing so. One was that Europe's caves were no friendlier a country for old men than Mexico's. Decades of extreme caving had taken a toll on his body, which happily accepted the occasional rest. More important was Klimchouk's conviction that supercaves required multigenerational efforts, which in turn needed some kind of inheritance mechanism.

Kasjan, a muscular, cheerful, sandy-haired man, now forty-nine, lives in Kiev. He was born in 1960 in western Ukraine in Sniatyn, a town of about ten thousand that dates from the twelfth century. As a child, he had Indiana Jones dreams, longing for travel to distant lands and grand adventures in them. The passing years tamed Kasjan's expectations a bit, but probably less than most people's. After secondary school, he attended Ukraine's Ivano-Frankivsk National Technical University of Oil and Gas, that country's equivalent of an institution like Texas A&M. He pursued a degree in geology, a field that seemed to offer some potential for Indiana Jones adventures.

Poking around in caves, of course, also dovetailed perfectly with that desire. Kasjan explored his first one in 1978 with a student speleological club from his hometown. Whatever "romantic" notions he held about caving were dispelled by the trip, which involved much of what cavers do best: digging and exploring dead ends. Later, he moved to the city of Poltava, also in Ukraine, and created a speleological club there, which still flourishes.

Today, when not leading expeditions, Kasjan works as an "industrial mountaineer," a relatively new, extraordinary profession that uses caving (and mountaineering) equipment and techniques to perform technical work on vertically extreme structures—skyscrapers, giant radio towers, oil rigs, and the like. Many of the same vertical techniques and equipment are used in both caving and mountaineering, but somehow the term "industrial caver" doesn't have quite the same ring.

Married twice, both times to cave explorers, Kasjan has two sons, Sergey and Denis, and a daughter, Anastasia. All are involved in speleology, the sons at the expeditionary level, his daughter more casually—so far. A three-time president of the Ukrainian Speleological Association, he is now editor-in-chief of the magazine *Svet* (Light), the association's official journal.

In major ways, Klimchouk and Kasjan perfectly complement each other. As both acknowledge, Klimchouk's caving passion is science-driven. Kasjan's has a more practical, engineering-inspired bent—somewhat like Bill Stone's, in fact. He derives immense satisfaction from solving practical, technical problems, such as those involving complex rope systems, huge vertical work, and intricate dives.

Kasjan first visited the Ortobalagan Valley in 1989 with a group from Poltava. He returned several times, and came back as leader of the Ukr.S.A.'s August 1999 expedition. An important goal was to explore those "windows" that had been identified earlier but not entered. Kasjan divided his team into two groups. One tried the lower window first, climbing from a nearly vertical wall into a passage that descended gradually for about 2,000 linear feet before ending in a closed chamber 1,600 feet beneath the surface. (Seen in profile on maps, this cave system looked very much like diagrams of the chambers and passageways in Egypt's great pyramids.)

That left the upper window. Klimchouk and Kasjan both knew that this window might be the magic portal. They also knew how unlikely that was. The numbers in caves here were as unforgiving as those in Mexico. Klimchouk, ever the scientist, quantified how things had played out over the years. Of hundreds of leads, 95 percent went nowhere; 4 percent yielded more depth and distance; only 1 percent produced substantial breakthroughs. Still, it was impossible not to feel excited. Klimchouk, and the scores of expeditionary cavers he recruited, had spent the better part of twenty years working up on the Arabika Massif, and by then all their hopes were focused on that hole in the wall in Krubera.

Alexey Zhdanovich, a young caver from the Ukrainian city of Uzhgorod, spearheaded this last push. Rappelling down into P43, he secured his descenders and hung there momentarily, looking through the window and into the tunnel as far as his light would penetrate. Hauling himself through the portal, he unclipped from the fixed rope and crawled into the darkness. This tunnel extended in the opposite direction from the one accessed through the lower window. It was also larger, big enough that Zhdanovich could crawl on his hands and knees, rather than slithering along on his belly.

He did not have to crawl for long. Before going 150 feet, he came to the lip of a pit in which a dropped stone took four seconds to hit the bottom. The shaft eventually proved to be 255 feet deep and unlocked the remainder of Krubera. This find was the critical turning point in Krubera's exploration. After Zhdanovich's lead through the upper window, other teams descended to almost 2,500 feet before the 1999 expedition ended. They stopped not because they ran out of cave but because they ran out of rope, supplies, and time.

When Kasjan and his crew finally emerged from Krubera, blinking in the painful sunlight, fellow expedition members presented them with bouquets of fresh flowers and mugs of good red wine. Everyone on the expedition knew that something very important had happened: Klimchouk's prediction that Krubera would turn out to be a true supercave had been validated. So had his faith in Yury Kasjan.

Breaking camp and derigging a cave is not usually the most joyful part of an expedition; it's more like the end of a party, when the lights come up and people realize that it's time to go home and that they will greet the next morning with a roaring hangover. This time, though, was different. Kasjan and his team members went about their mopping-up duties with unusual good cheer, buoyed by the knowledge that the cave was still going. It took a lot to make these people of endless winters and dark history smile, but what they had just found deep in Krubera was more than enough.

THIRTY-SEVEN

THINGS THEN WENT SO QUICKLY THAT, later, Klimchouk confessed to feeling a bit mystified, because things rarely went quickly in supercave exploration.

He sent another Yury Kasjan–led expedition, composed solely of Ukrainian cavers, in August 2000. That group descended to 3,888 feet and left the cave rigged for another group, one that, for the first time, included non-Ukrainians. The Spanish caver Sergio García-Dils (who had helped rescue Alexander Kabanikhin) had been imploring Klimchouk for a Krubera expedition spot. So had the prominent French caver Bernard Tourte. Yury Kasjan obligingly agreed to lead this group as well, despite having already spent the better part of August underground.

On August 31, García-Dils, with other Spanish and French cavers, met Kasjan in the Georgian city of Sochi, a Black Sea resort town with an international airport (and coincidentally, the scheduled site of the 2014 Olympic Winter Games). Most of the fighting between Abkhazia and Georgia had ended, but not the centuries-old animosity that divided the two populations. Both sides were guarding their borders, as the Russians were their own.

Unwilling to pay the exorbitant bribes ($300 per caver) demanded by corrupt Georgian border guards, they eventually retained the services of a "fixer" named Vatek. He arranged smaller bribes to Russian guards at a different crossing. There, the Euros waded the Psou River, which separates Russia from Abkhazia, on a pitch-dark night. On the other side, they walked a long way through open fields, which Vatek had neglected to tell them were thickly sown with land mines. They were lucky. All were intact the next morning when Vatek showed up in a van loaded with their caving gear.

The next day, the Arabika Massif proved to have more objective dangers than storms and avalanches. Yury Kasjan and his team were putting the finishing touches on their base camp when a group of AK-47-toting Abkhazians showed up. Cavers had been working in the Ortobalagan for twenty years by then, so most of the locals were accustomed to seeing people dressed in brightly colored suits disappear into the ground for weeks at a time. The cavers hoped that these soldiers would be similarly nonchalant about their expedition. Fortune was with them; the squad had dropped by to say hello and see how things were going. At one point they handed over their rifles so that the visitors could do a little full-auto plinking.

Kasjan's teams explored down to 4,600 feet. At that point, progress ended at a tight squeeze in a meander with no airflow. Ascending a shaft after the expedition's last drop to the bottom, Yury Kasjan found another window, this one at about the 4,000-foot level. Here air *was* moving, and there was a special feel to the place. Klimchouk was not the only one with great cave intuition. Something about that window spoke to Kasjan; he could *smell* depth there.

Tantalized, Klimchouk and his cavers decided not to wait for the next summer and organized a winter expedition. In one important way, winter was better: with everything frozen on top, the danger of flooding down low was virtually eliminated. Set against that were the difficulties of getting up onto the Arabika Massif in winter, when many feet of snow and ice required the establishment of a mountaineering base camp at the cave's mouth. Those winter conditions also made it harder to bring supplies in and any injured people out.

Alexander Klimchouk once again oversaw organization of the expedition, and the redoubtable Yury Kasjan led it. This time, though, the eleven-person team included members of a Moscow-based group called the CAVEX (Cave Exploration) Society, which Klimchouk had helped create in 1998 as a kind of daughter group under the Ukr.S.A. umbrella—his fondness for affiliation at work again. Klimchouk had seen a need to improve working relations between Moscow cavers and the Ukr.S.A. His good intentions, alas, would produce a classic demonstration of that old saying about no good deeds going unpunished and CAVEX would have, like Dr. Frankenstein's creation, a catastrophic effect on his life.

Klimchouk's son Oleg, then twenty-four, was a CAVEX member and had lobbied his father heavily to include the Muscovites in the winter expedition. Klimchouk had no qualms about Oleg. Given his early start in caving and his love for the activity, Oleg was competent in all the required disciplines. Even so, at first Klimchouk hesitated, for reasons that had nothing to do with concerns about Oleg's capabilities.

This expedition could produce a once-in-a-century discovery. The Russians had not contributed to previous Krubera explorations, so their unearned opportunity to "skim the cream" (the Ukrainian equivalent of "scooping booty") would be resented by many in the Ukr.S.A. But Klimchouk was the kind of man who listened hard for the better angels of his nature, and he decided to let them come. The Moscow guys were young, strong, and ambitious cavers and, after all, they were also technically Ukr.S.A. members. Why not give them the privilege of taking part in what might be a historic event? Why not encourage, in this way, international cooperation in cave explorations? Klimchouk's intentions were good, but they ended up paving the way to a hellish confrontation not only between his association and CAVEX but also between him and Oleg.

Initially, team leader Kasjan was also reluctant to accept the Russians, but Klimchouk persuaded him. They both made it clear to the Moscow contingent, however, that they were coming along as part of the Ukr.S.A. expedition, not as a CAVEX group. The eager young cavers stated that they perfectly understood the situation and happily agreed to abide by Klimchouk's terms.

The expedition departed from Kiev on Christmas Day 2000. Two days later, a helicopter dropped them and their supplies into deep snow at the base camp area. Whether in Oaxaca or Arabika, rigging is always the first order of business. In just two more days, they'd established Camp 1 at 1,600 feet deep, at the bottom of a 450-foot shaft with a disconcerting tendency to spit rockfall. Staying there was rather like being an infantry platoon under mortar fire.

Unfazed, the cavers soldiered on, working with the efficiency and cooperation typical of Ukr.S.A. teams. Four duos toiled around the clock. Their ultimate mission was to explore the window that Yury Kasjan had found in August. The honor of going first went to the expedition's youngest member, Anatoli Povyakalo, who had recently celebrated his eighteenth birthday. Povyakalo succeeded in pushing through the window and down to 4,756 feet. Increased air and water flows suggested that they were within striking distance of the then-deepest cave in the world, 5,354-foot deep Lamprechtsofen, in Austria. Late at night on January 4, Moscow caver Ilya Zharkov and a companion broke that record, descending to a huge chamber at 5,510 feet, where a major boulder choke stopped them.

Four more explorers took over the point, hoping to go even deeper, but there was no way around the choke. Expedition members named the vast new room the Chamber of Soviet Speleologists. Unlike extreme mountaineering expeditions, which typically place only a few climbers from a large group on the summit, every member of the 2000–01 Krubera probe got below 5,000 feet, and nine went through a new passage all the way to the expedition's deepest penetration, nearly 1,710 meters, or 5,609 feet. Krubera was now officially the deepest cave on earth, which certainly was cause for celebration, but everyone expected that it would go much deeper, perhaps another 1,000 feet or more.

The hazardous work of descending more than a mile into Krubera went off without a hitch, but not so the team's descent from Arabika. They waited anxiously for a scheduled helicopter to pick them up on January 11. Bad weather grounded it at sea level that day, and then worsened. Forecasters predicted high winds, heavy snow, and near-zero visibility for at least three days. The team decided to walk down off the Arabika Massif. They were about three miles from the timberline, and their route took them through terrain where the avalanche hazard was high. They'd made it a bit more than halfway when a large avalanche let go from a mountain face right in front of them. It caught and buried the young caver who had so recently celebrated his eighteenth birthday, Anatoli Povyakalo. Proving that he was the luckiest as well as the youngest, Povyakalo survived, thanks to the fast action of his mates, who dug him out before he suffocated. The remainder of their hike out passed uneventfully, and on January 16 an old gray helicopter, taking advantage of a brief weather break, evacuated their remaining gear from base camp.

BACK IN KIEV, ALEXANDER KLIMCHOUK HAD spent the first week of January 2001 anxiously awaiting news from Kasjan. Klimchouk knew full well that this expedition might bring to fruition decades of grueling, dangerous work. One day after another passed, until he was literally pacing the floor. Then, on January 7, the phone rang. It was Denis Provalov, calling from Arabika on his satellite phone.

"Well, Alexander, I just came to the surface alone. The weather has been a bit wintry, but everything has proceeded pretty well," Provalov said matter-of-factly. Imagining Klimchouk in a lather of impatience, Provalov could barely suppress his laughter.

"*But . . . ?*" Klimchouk broke in. "What happened at the bottom?"

"Well, Alexander, perhaps you had better sit down," Provalov said. "Because we broke the world record here. Krubera is now the deepest cave on earth."

Then Klimchouk did sit down. The announcement literally left him breathless. This moment was the culmination of nearly thirty years of work, beginning with the discovery of Kilsi Cave in central Asia and continuing through decades of organizing enormous expeditions and taking immense risks year after year. The long dream had become, at last, a reality. It took a while for the fact to sink in. And although Klimchouk had not made it to the bottom on this particular trip, he *felt* as if he was there, at the bottom of Krubera, the new deepest cave in the world. He rightly considered it the result of his work of decades. The glow lasted for weeks.

According to Klimchouk, he was not the only person Provalov notified with the sat phone. Provalov also called Moscow media with word that "Russian cavers have established the new world record." It was in Klimchouk's view "already the beginning of [a] misinformation campaign."

On January 18, a huge crowd jammed the Kiev railway station, awaiting the arrival of the Sochi train. People came bearing flowers, vodka, champagne, and banners. When the train pulled in, an orchestra launched into a victory march. Television crews and newspaper reporters jostled for position, shouting questions. The expedition members, coming one by one from the train onto the platform, were immediately hoisted onto the shoulders of celebrants and paraded around like heroes.

The new world depth record generated significant coverage in Europe and the United Kingdom, where the general public follows activities like mountaineering, caving, climbing, and diving more closely than do Americans. Even so, the publicity was nothing like that lavished on polar discoverers and summiters of 8,000-meter peaks.

Earth scientists and speleologists certainly understood how significant the Krubera descent really was, and what it portended. Krubera's penetration had exceeded the previous record by 250 feet in one fell swoop, while advances for the depth record during the previous three decades had averaged less than 50 feet per year. In addition, it was the first time in the history of geology that the deepest cave in the world had been established outside of western Europe. (For a correlative, imagine that the new highest mountain on earth was suddenly discovered in Chile.) During the subsequent month, congratulations by the hundreds poured in from universities, caving societies, academic societies, and scientific organizations all over the globe. It was infinitely gratifying for Klimchouk that people all around the world shared his joy and excitement.

Klimchouk and his team enjoyed the thrill of victory for weeks. But after a while, with the champagne drunk and the exultation quieted, he began to experience the angst that almost invariably sets in after the accomplishment of a great goal. He understood, as transformers do, that discovery is a two-edged sword, cause at once for celebration over great accomplishment and sadness at a long journey's end. What could he possibly do as an encore, now that his team had discovered the new deepest cave in the world?

It didn't take Klimchouk long to decide. In February 2001, he established a new project for the Ukr.S.A. and CAVEX: to discover the first 2,000-meter (6,562-foot) deep cave on earth. Despite Klimchouk's accomplishments, many knowledgeable cavers and geologists scoffed. Before the Krubera exploration, explorers had taken twenty-five years to add about 1,000 feet to the depth record. Now Klimchouk thought he could almost repeat that feat in only a few years? Not likely. But Klimchouk had surprised the scoffers before.

Thus was born the Call of the Abyss Project.

THIRTY-EIGHT

GREAT DISCOVERIES HAVE A WAY OF generating great tragedies and controversies. After Amundsen's fast, light expedition used sled dogs and skis to discover the South Pole, some (especially the second-place Brits) accused him of dishonorably running a "race," rather than conducting a serious scientific expedition. When, during his own South Pole quest, Robert Falcon Scott's entire team perished after being bested by Amundsen, others questioned Scott's leadership and judgment. Before Hillary and Norgay were even off the mountain after summiting Everest, a divisive question was being asked: *Which man got there first?* True to form, things began to go sour not long after Krubera was proven to be the deepest cave on earth. Sadly, the worst of it was between Alexander Klimchouk and his son Oleg.

Alexander Klimchouk started caving young, at around eleven. If he had had a caving father, he probably would have started even earlier, as did his own sons. Some sons almost reflexively rebel against their fathers' predilections, but not Oleg. He took to caving quickly and passionately, which brought great joy to Alexander. What richer reward does life bring a father than to be joined by one he loves most in doing a thing he loves most?

By then twenty-five and working in the same kind of industrial mountaineering as Yury Kasjan, Oleg had made a move typical of maturing young men, relocating from quiet places to where the action is. He spent a lot of his time in the former Soviet Union's happening place, Moscow. That city was also the home base for CAVEX, which Oleg enthusiastically joined.

Soon after returning from the cave, Oleg asked his father's permission for himself and CAVEX to seek commercial sponsors and raise funds using the world record in Krubera as their leverage. Against his better judgment, Klimchouk agreed, with restrictions. Please, he said, remember that CAVEX is there only as part of the Ukrainian Speleological Association, and that the Ukrainians had done all the years of advance work that made the record possible.

No problem, Oleg and the Muscovites assured Klimchouk. But problems arose almost immediately. CAVEX members began claiming in interviews and articles that the new deepest cave in the world had been discovered by the CAVEX team. The Russian media have no love for things Ukrainian, so several major Russian television channels' follow-up reports portrayed the CAVEX team as heroes and the Ukrainians, when they were mentioned at all, as mere supporters. This would have been bad enough had the misinformation been broadcast exclusively in Russia, but those stations were also dominant throughout Ukraine and became the basis for international coverage, as well. The Ukrainians were shocked and felt disgraced in the eyes of the world.

"They started to go mad about the idea of superiority and fame," Klimchouk said of CAVEX. "The world record clearly made their minds go awry. They started to develop plans to promote CAVEX as a super-extreme team and make a living on it."

Klimchouk was distressed, but what could he do? That horse was already out of the barn and far down the road. His lone voice and those of a few Ukrainians were drowned out by the thunder of the international media. He took some small comfort from the fact that modern news stories have a twenty-four-hour shelf life, if that. Soon enough, the stories would ebb. Things would return to normal, and he and his teams could go back to the business of exploring Krubera.

Or so he assumed, until one day in the spring of 2001. The media interest had faded, but then Oleg approached his father with a truly bizarre plan. Representing CAVEX, Oleg proposed that his father should force Yury Kasjan and the entire Ukr.S.A. out of Krubera and let CAVEX take it from there. The young Turks—or Russians, actually—in Moscow had been plotting a regime change.

Klimchouk was horrified. It was as though Oleg had asked him to cut off an arm. The request, and Klimchouk's angry refusal, sparked a severe conflict between the two, opening a breach that remains to this day. To understand the depths of Klimchouk's distress, it helps to know that he, the gentlest of men, was so angered that he "was about to beat" Oleg. That's hyperbolic; Klimchouk never struck Oleg, or really even came close. But he was undeniably furious. For his part, the son, realizing what a terrible mistake he had made, backed off and steered clear of Alexander for some time.

It was a tragic example of the law of unintended consequences. With the best of intentions, Alexander Klimchouk had introduced his son early on to the thing that had been the guiding force of his own life, cave exploration. And though he loved both his son and cave exploration without reservation, somehow now the two had conspired to cause him the greatest pain he had ever suffered. The breach between father and son was something else as well, eerily reminiscent of the divide that had opened between Bill and Pat Stone.

Nevertheless, Klimchouk chose to at least partly give the Muscovites the benefit of the doubt, persuading himself to write Oleg's gaffe off to youth and the influence of young Russians with stars in their eyes. Thinking that direct communication between himself and the Russians would restore them to their senses and bring dignity back to the exploration of Krubera, Klimchouk invited them to take part in his upcoming expedition to Turkey's Aladaglar Massif, which had significant supercave potential. He personally led the twenty-three-member expedition, which included a number of CAVEX members.

During a month in the Turkish mountains, Klimchouk did achieve a better understanding of the CAVEX people, but it only confirmed his worst fears. The limelight was and had always been, for Klimchouk, far down a long list of more important objectives; doing serious science, protecting the environment, and team safety, to name just three. He concluded that the Russians were apparently obsessed with fame, with ginning up accolades from the technically ignorant media for even routine explorations. He felt that his Russian invitees poisoned the month-long expedition, perpetrating what he interpreted as betrayals to garner inordinate credit for themselves.

Worst of all, in fighting with CAVEX, Klimchouk was fighting with his son, perhaps giving that frayed relationship the last tug needed to rip it apart forever.

So dismayed was Klimchouk by what he perceived as the CAVEX team's duplicity that he elected to stay away from his beloved Arabika in 2002 and 2003, years in which CAVEX teams were there. Instead, he and Yury Kasjan led expeditions to the Aladaglar Massif in both years, monitoring the CAVEX efforts from afar. Klimchouk certainly resented the Russians' cheeky invasion, but he also worried about the younger, less experienced teams. He feared that they were badly organized, included too many inexperienced cavers, took too many risks, and were, in sum, a recipe for disaster.

THE 2003 CAVEX EXPEDITION KICKED OFF on July 29. The base camp was quickly established. The long-haul travelers shrugged off their jet lag, adjusted their watches, and everybody got used to sleeping on (and in) the ground once again.

By then, the world depth record had changed yet again. On January 12, 2003, an expedition had descended to 5,685 feet in France's Mirolda Cave, besting Krubera by 74 feet. And there was always the specter of Bill Stone's Mexican exploration at Sistema Cheve, which, he had unabashedly announced, would prove to be the deepest cave on earth.

Several unexplored windows and passages led out from the vast Chamber of Soviet Speleologists, the vast room near the bottom of Krubera. Others existed on the way to it. Of highest interest, though, was that small sump located 4,700 feet deep at the bottom of a 36-foot drop called P11. Following protocol, the cavers had bypassed these "possibles" as long as they could keep going through main passage. Now, halted for good at 5,609 feet, it was time to go back and dive that sump.

By early August, teams had rigged most of the cave, created a camp at 3,985 feet deep, and installed a telephone line. Things had been moving along so beautifully that everyone could have been forgiven for thinking that this expedition was blessed with a guardian angel. That lasted until a squad of soldiers armed with AK-47s showed up.

THIRTY-NINE

ONCE AGAIN, THEY WERE ABKHAZIANS. Dropping in unannounced, they stated their intention to stay for "a few days." Perhaps because of the Europeans on this team, or for some other, unknown reason, these soldiers seemed less friendly than the ones who had dropped in on the August 2000 expedition. This was part of their training, they explained, and in addition, they wanted the cave explorers to feel safe and secure, which, they had no doubt, the presence of stalwart Abkhazian troops would ensure. In fact, the cavers had shrewdly anticipated such an event and made sure that their camp was flying two flags—one Abkhazian, the other white. The soldiers hung around, smoking and watching, but did not interfere.

The cavers went back to work, and on August 14, Oleg Klimchouk entered the world's deepest sump. Around the world in Mexico, Bill Stone, Rick Stanton, and Jason Mallinson had demonstrated that all the immense preparation required to make a supercave dive might produce only a few extra yards of passage and depth; or it just might unlock miles, as Stone and Barbara am

Ende's had in Huautla in 1994. It was always a crapshoot, and this time the dice came up snake eyes. The sump Oleg Klimchouk dove was almost comically anticlimactic: only 13 feet long and 6 feet deep, it quickly narrowed down to a hole the size of a basketball. Beyond, Klimchouk could see no bottom, but with the poor visibility, that could have meant anything from 10 feet to 1,000.

Klimchouk's dive *was* the deepest ever done in a cave, but it did not take him deeper than the French had gone in Mirolda. Thus it was worth not even a footnote in the record books—*unless* it opened the way to go much deeper in Krubera. Accordingly, teams brought down more air tanks, as well as tools for enlarging the hole at the sump's end.

On the evening of August 18, Oleg Klimchouk and Denis Provalov dove the sump together, slipping through the newly enlarged portal into the big chamber beyond. Finding dry walking passages at its end, they explored on foot until they came to a large waterfall rushing over a cliff. With neither rope nor hardware to descend farther, they returned through the sump. That second dive marked the 2003 expedition's deepest penetration, which broke no records. It was time to begin the long ordeal of derigging.

Several teams, including Sergio García-Dils's, were working deep in the cave when, early in the morning on August 22, a storm struck. Above ground, wind and cloudburst rains damaged tents and scattered gear. According to accounts published later by the expedition leaders, the steady, hard rain produced torrents of water that began flooding the cave. Worse, they were then struck by an earthquake. For the record, Alexander Klimchouk, though he was not on the scene, remained skeptical about such dramatics. To his way of thinking, the "flood" was no more than routine water inflow for that time of year. Nor, he felt sure, had there been an earthquake. He attributed the tremors to a rockfall deep in Krubera or a nearby cave. Klimchouk concluded that the 2003 cavers' relative inexperience had led them to describe as catastrophic events that would have left more seasoned veterans unfazed. Lending credence to Klimchouk's theory, Sergio García-Dils later wrote that the "earthquake" felt by those inside the cave went unnoticed by those on the surface.

The storm did, unquestionably, bring heavy lightning. One bolt struck the communications center on the surface, sending a charge down through the telephone line. Ilya Zharkov happened to be talking on the phone at that moment. The receiver jumped out of his hand, and he flew ten feet through the air. Despite his thick caving gloves, Zharkov's hand was badly burned. He was probably saved from outright electrocution by the thick rubber soles of his caving boots.

What with the storm, possible earthquake, flood, and lightning, García-Dils and the others imagined that before long they might start hearing apocalyptic trumpets. The lightning storm abated eventually, but the rain continued, and the flooding in the cave really did get serious. Two separate pairs of cavers found themselves trapped in meanders between the 500-Meter Camp and the one down at 4,000 feet. One pair survived, barely, by climbing to the ceiling and clinging to the wall of their particular passage, which had flooded almost completely. The other fled desperately from an onrushing wall of water, jumping out and to the side of a tight passage just in time to avoid being washed away. Shades of Indiana Jones.

No one was killed by the flood, quake, or lightning. By the morning of August 23, the high water was receding and the derigging resumed. At about 11:00 A.M., Alexander Kabanikhin entered the cave and began working his way down, rappelling successive pitches. He made his way through a 360-foot meander called Ulybka, which, translated, means "Smile." He came to the edge of the Big Cascade and rappelled over its lip. At the first rebelay anchor, he transferred his descender to the second rope, leaned back, and fell away into the darkness.

FORTY

WITH KABANIKHIN SECURED IN THE 500-METER Camp, discussions began about how to conduct the rescue. Cave rescues are always difficult. For a variety of reasons, this one would be horrific. Sergio García-Dils was the only trained search-and-rescue expert on the whole team. In addition, those who would have to perform the rescue were already worn out from weeks of grueling labor. They had not brought with them even basic rescue tools: a litter, pulleys, winches, and other hardware. Everything would have to be accomplished from within Abkhazia, a region that was unknown to most of the world and that was suffering shortages of just about everything. And as if all that were not enough, every discussion, decision, and movement was further complicated because cavers from seven countries were present and many spoke only their own native language. Finally, while the flooding had eased, the waters were still not down to their prestorm levels, and the essential telephone communication system had not yet been repaired.

Shortly before noon, the team decided that the experienced French caver

Bernard Tourte would remain with Kabanikhin while Sergio García-Dils sprint-climbed to the surface to organize a rescue. The Spaniard arrived on top at 1:00 P.M., only to learn that the telephone system had been repaired and news of the accident had preceded him. Regardless, given his experience in search and rescue, García-Dils became the "incident commander."

The closest possible source of rescue equipment was the Russian Ministry of Emergency Situations in Sochi. Requesting help from Georgia-hating Russians might be asking for trouble, but what could they do? That afternoon, the expedition contacted the Russians in Sochi and formally sought assistance. Meanwhile, García-Dils canvassed the base camp, gathering every tool that might be of use during the rescue. He also delegated a team to start making explosive charges to enlarge tight squeezes for a litter's passage.

The Russians promised to send a helicopter with equipment, supplies, and more people. That evening García-Dils gave a crash course in cave-rescue techniques to those in the group who had never participated in one. Extricating Kabanikhin through so much vertical terrain was going to be complicated by the weight of the loaded litter, the convoluted passages, and the difficulty of creating complex anchored pulley systems, among other impediments. At some points, applying the same principle used in elevators, human "counterweights" would be required to raise the litter. At others, it would have to be pushed and pulled through meanders big enough to admit it but not large enough to stand in. A purpose-built system for this kind of thing, called a confined space rescue system, includes dozens of specialized devices and costs more than $5,000. This rescue would have to be accomplished with six well-used pulleys García-Dils had managed to scrounge during his campwide search, unless the Russians brought better gear—but when did Russians ever have better gear?

At three the next morning, the cavalry choppered in: eleven fresh bodies, medical supplies, and a litter. It was not what the cavers had been hoping for. Cave-rescue litters are made from tough, thick plastic that retains some flexibility, allowing the litter to be strapped around the victim like a cocoon. They are also light, weighing only ten to fifteen pounds. The Russians had brought a litter designed for helicopter rescues, a rigid, fifty-pound metal basket. Riggers had spent the rest of the night putting special anchors in place, and shortly before noon a rescue team started down with the litter.

While so much had been transpiring on the surface, others had been working feverishly in the cave to enlarge passages. The Mozambique Meander, for example, was several hundred feet long and averaged about 2 feet wide. Most people had to turn sideways to navigate it. By 12:30 P.M., workers had hammered and blasted enough openings that cavers could wrestle the empty litter through, though to pass several places it had to be disassembled into two halves. Four hours later, the litter arrived at the 500-Meter Camp, where Kabanikhin had been languishing, his suffering eased by injections of tramadol, a powerful narcotic-like analgesic.

It took three hours to stabilize Kabanikhin and secure him in the litter. The load totaled almost 250 pounds. At about 7:00 P.M., he was lifted from the floor of the 500-Meter Camp and began the long, hard trip to the light. Rescuers hoisted the litter up through the deep shafts and wrestled it through squeeze after squeeze. By 1:30 A.M. on August 25, they reached a prearranged bivouac spot, about halfway to the surface. There, rescuers and victim rested while others kept working to widen one of the worst passages of all, a vertical section that rose from 250 feet deep to about 160 feet. A rugged Ukrainian named Alexey Karpechenko—nicknamed "Brick" because of his toughness—and his partner, Nikoley Solovyev, had been working almost nonstop in this section. It took 120 explosive charges and a gasoline-powered jackhammer to get the job done.

At 7:30 P.M. on August 25, rescuers resumed hauling the litter up; the work continued into the next morning. At 4:00 A.M. on August 26, sixty-four hours after his accident, Alexander Kabanikhin emerged, Lazarus-like, from the mouth of Krubera into fresh air he must surely have thought he would never breathe again.

KABANIKHIN WAS SUCCESSFULLY EVACUATED AND SPENT many weeks in the hospital. Amazingly, he suffered no lasting impairments from the catastrophic accident that could so easily have ended his life.

The accident was more than enough to end the 2003 exploration of Krubera, which, as Klimchouk had feared, had brought tragedy without producing any triumph. For all their hard work, the explorers had not succeeded in going deeper in than 5,609 feet, Krubera's known depth at the expedition's beginning. Alexander Klimchouk knew that cavers just as skilled and determined as his teams were working elsewhere, and he was particularly cognizant of the work being done in Mexico by Bill Stone and the United States Deep Caving Team.

Stone had never hesitated to proclaim to the world that Cheve had a proven, dye-traced potential to be at least 2,500 meters (8,202 feet) deep and that he, Bill Stone, would penetrate to that depth, even if it meant spending more than a month underground and diving far enough to effectively separate from the main expedition. Klimchouk had met Stone several times during visits to the States and had been impressed by the force of his personality and the magnitude of his accomplishments. Klimchouk knew that Stone had been on almost fifty Mexican expeditions and that he was brilliant, experienced, and driven. He knew that the National Geographic Society was one of Stone's sponsors, and if there was one thing you could say about that organization's executives, it was that they did not back losers. If anyone, anywhere, other than Klimchouk's teams was likely to make a major breakthrough in a supercave, the odds were with Bill Stone in Mexico. Klimchouk knew that Stone, with ample sponsor support, legions of expert explorers, the most advanced diving technology, and white-hot ambition, planned a return to Cheve in 2004.

You could interchange "Stone" and "Klimchouk" often in the paragraph above and produce an accurate picture of the American caver's view half a world away. Stone knew that Krubera, like Cheve, had the dye-trace-proven potential to go at least 8,000 feet deep. He knew that Klimchouk, like himself, was a dedicated, mission-focused scientist with tremendous resources who would not spend a minute in Arabika unless he saw ultimate potential there. Klimchouk had at his disposal even more, and more disciplined, expert cavers, many of whom had been exploring the depths since early childhood. Stone's sponsor, the National Geographic Society, would be supporting an all-out effort by Klimchouk's Ukrainian Speleological Association the next year. This gave the already ultramotivated Stone more incentive, if any were needed, to make his own all-out attempt in Mexico in 2004.

Thus the stage was finally set for an epic race not just to find the deepest cave on earth but also to make the last great terrestrial discovery. Nothing like it had been seen since 1911–12, when Amundsen and Scott had raced across the barren Antarctic wastes toward the South Pole.

Now 2004, it appeared, could well be that same kind of year.

PART THREE
GAME OVER

Either the air is drunk or the hobgoblin is zealous.

—Russian poet Leonid Filatov, quoted by Emil Vash

FORTY-ONE

ON OCTOBER 28, 2003, about six months after Bill Stone's thwarted expedition decamped, a Mexican rancher named Pedro Pérez was walking to his home in the village of Santa Ana Cuauhtémoc, population 783, in the same area as Cheve Cave. Santa Ana occupies a small saddle 4,249 feet high, between two mountains of the Sierra Juárez. Pérez's route took him close to the San Miguel River, which flows through the bottom of a deep canyon called the Star Gorge. Pérez had lived here all his life, and knew the sounds of the San Miguel almost as well as he knew his wife's voice in all her moods. Heavy rains this year had the river roaring like never before.

When Pérez reached the riverbank, he stood and stared. The San Miguel was flowing at about twenty times its normal volume, a churning, boiling torrent of chocolate-brown water and creamy foam. Suddenly, without warning, the river simply disappeared into a huge maw in its bed. The hole was almost 100 feet in diameter and about 40 deep. Where the river went after pouring into the huge hole, Pérez could not imagine.

News of the new sinkhole spread quickly through the Cuicatec community and from there to Mexican cavers, who knew that Bill Stone would be interested in it. He was, so much so that it became the primary target of his USDCT 2004 Cheve expedition. He thought it possible that if they could enter the sinkhole during the dry season, it might prove to be a "back door" to Cheve, far above. It was the same kind of hope he had held for Peña Colorada, exactly twenty year earlier.

His expedition had three targets. The first was the new sinkhole. The second was an existing cave upstream from the sinkhole called Star Gorge Cave. The third was a large "recon zone"—an unexplored area that Stone wanted to learn more about—high up in the nearby Aguacate Cloud Forest, a forbidding area of pits, snakes, and cacti at about 8,000 feet on a mountainside above a nearby village called San Francisco Chapulapa.

Stone led an expedition down to Mexico in 2004. He and Andi Hunter, the strong and striking Alaskan outdoorswoman, were now officially a couple; in addition to them, the core team had five other members. A former race-car driver who now lived in Pennsylvania's Amish country and was a manager for the New Holland tractor company, tall, talkative David Kohuth (Ko-HOOTH) was an experienced, indefatigable caver. Gregg Clemmer, an award-winning historian and a lifelong caver, built like a fireplug, hailed from Germantown, Maryland. Wiry, affable Ohioan John Kerr, the technowizard who had been such an invaluable part of the 2003 expedition, accompanied the team as well. Jim Brown, quirky and introverted but also one of the world's top cave divers, came along to penetrate any potential sumps. Finally there was newcomer Ryan Tietz, handsome as a film star and tough as nails, who had just graduated from Texas A&M. Ultimately, the expedition would include thirty-eight cavers from eight countries.

The 2004 expedition left Austin, Texas, on February 12 in two overstuffed red trucks and David Kohuth's Jeep Grand Cherokee. After two days of nonstop travel, the caravan finally pulled off the road at 2:00 A.M., and the cavers rolled out their sleeping bags and got some much-needed rest. The next morning they were awakened by the sweet smell of sugar cooking in a nearby mill. A few hours later, they rolled into Cuicatlán, a town of about fifteen thousand guarded by red, thousand-foot-high cliffs.

The next day, Stone and Hunter met with local Cuicatlán officials to secure permission for their expedition. That done, they headed for a popsicle stand—even in February, Cuicatlán was baking by afternoon. As they crossed the street, an armed policeman in blue fatigues, eyes hidden by dark aviator glasses, halted them. Stone and Hunter braced themselves for a third-world imbroglio. The man in blue, however, broke into a broad smile. Introducing himself as the chief of police, he whipped out the February 2004 Spanish issue of *National Geographic* magazine, which featured Bill Stone's article about the 2003 Cheve expedition. *El Jefe* wanted an autograph, which Stone gladly provided.

The journey to their caving destination resembled passing through a series of successively smaller air locks. From Cuicatlán, the team traveled six miles northwest and thousands of feet higher to Concepción Pápalo, a village of several thousand and home to another local authority with jurisdiction over the land where Cheve Cave's entrance was located. They then continued northwest over the top of the mountain on the western flank of which Concepción Pápalo perched and headed down the other side to Santa Ana Cuauhtémoc. It was near there, the previous October, that Pedro Pérez had first spotted the huge sinkhole.

The cavers hacked space for their main base camp out of the jungle near the Star Gorge sinkhole. Over the next few days, pack trains of mules, horses, hired locals, and the cavers themselves hauled thousands of pounds of food, clothing, equipment, and other supplies.

Stone and the team quickly found the gigantic hole in the now virtually dry riverbed. During the rainy season, it would have been, as Pérez had asserted, big enough to swallow the river. The hole resembled a pit such as might have been excavated by one of those huge "bunker buster" bombs—except that this one had no discernible bottom.

Doughty Gregg Clemmer free-climbed almost 20 feet down the steep, rocky wall on one side of the hole and could see only more gaping space. The "wall," though, consisted of very large boulders precariously balanced. Knowing that anything—or nothing—could cause a deadly collapse, Clemmer wisely decided to climb back out.

Before the team could explore further, those teetering, truck-sized boulders had to be removed. For most expeditions that would have been game over right there, but not for one of Stone's. He had brought along two five-ton-capacity electric winches made by the Warn Company, as well as fourteen GM Ovonic electric-car batteries to power the winches. Once attached to solid rock walls or giant trees, the winches could extract boulders from the riverbed like a dentist pulling huge teeth.

There was serious work ahead, but also cause for celebration, given the size of that sinkhole. Those who might have wondered if their leader ever cracked a smile, let alone a laugh, had their answer that night. Fueled by excitement and good spirits of both kinds, the team had a rousing sing-along in the kitchen tent after dinner. Stone, wearing what looked like a turquoise turban and a huge smile, sang and strummed accompaniment on his guitar, despite a right index finger swathed in bandages. Andi Hunter kept the lyrics flowing with a sheaf of sheet music she had brought. Even Jim Brown, hard of hearing and slightly bemused, joined in for a chorus or two.

On February 17, thinking it wise to start by following the path of least resistance, Stone put the whole team to work on Star Gorge Cave. By the time they were ready to leave camp, at midday, four more explorers had arrived. Two were Americans; the others were a couple, Jan Matthesius and Pauline Barendse, Stone expedition regulars who had come all the way from the Netherlands. Veteran cavers all, they strode into camp with the bright eyes and high energy that typify the beginning of every expedition. They could not have suspected that the coming weeks would present challenges and agonies unlike anything any of them had encountered before.

FORTY-TWO

SOME CAVES ARE MORE WELCOMING THAN others, and Star Gorge Cave ranked near the bottom. For starters, thousands of vampire bats roosted on the ceiling of its entrance chamber. Though they hung upside down, these bats managed, by way of some very acrobatic elimination, to deposit daily many gallons of bloody liquid excrement on the cave floor below. There it collected into stinking, fetid, germ-ridden ponds about the consistency of yogurt. The bat guano was no laughing matter. Following a 2001 reconnaissance in Star Gorge, caver Marcin Gala contracted histoplasmosis, a fungal infection in the lungs that can fatally destroy one's respiratory system if not treated promptly. It required two full weeks in the hospital and copious doses of powerful antifungal medications to save him.

Having passed that gauntlet, the cavers worked their way 600 vertical feet to the bottom of Star Gorge Cave and began digging. There was no airflow through this cave, and the atmosphere down deep quickly became foul. Back in Maryland, the bearded, photogenic scientist Bart Hogan, whose inventive skills equaled John Kerr's, had created little blower units and attached them to long hoses. These he'd then connected to a Bill Stone rebreather unit. The setup allowed one blower to remove stale, carbon-dioxide-laden air while the other pumped fresh oxygen in.

This was cave digging at its worst—or best, for those who relished such a challenge. It took an hour of hiking from camp and another of descending into the cave just to reach the end where digging was taking place. During the expedition's first days, even the hiking could be treacherous, taking the cavers down steep, rain-slicked paths laced with slippery roots. But as the days passed and the trails dried, the cavers started looking forward to these walks under blue sky and good sun. The scenery was spectacular up there, emerald, boulder-strewn meadows interspersed with fields of coffee plants and stands of bamboo. The altitude provided spectacular views of the dramatic mountains that surrounded them.

Another hour of rappelling and downclimbing brought cavers to the dig, where they would spend the next seven, highly unpleasant, hours. Their excavating system resembled a giant human centipede. Deep in the hole, one caver hacked and chopped away at the concrete-hard earth, either filling a bucket beside her or pushing the dirt back to a partner, who put it in a bucket. That was passed back from caver to caver, until those farthest to the rear pulled the buckets out and dumped their contents in the open cave. Filled with sand and gravel, each bucket weighed about fifty pounds. Twenty buckets constituted a good day's haul—about a ton of material, in other words, which produced a gain of 6 to 12 feet, depending on the consistency of the dirt.

By the time their shift ended, the workers emerged from the dark of the cave into the dark of night, their hike back up to camp illuminated by stars and moon, when clouds did not obscure them. After three days of digging without breakthroughs in Star Gorge Cave, Bill Stone moved some of the troops to the big upstream sinkhole, the one first observed by Pedro Pérez. By now, the riverbed was dry and water no longer presented a hazard. A team led by Gregg Clemmer dug down 4 feet through the sinkhole's sandy bottom, about 40 feet below the riverbed. Just after sunset, Clemmer, deepest in the hole, poked his custom-designed titanium digging tool (it resembled a crowbar) through the dirt into . . . *nothing.* He quickly bashed an opening 2 feet in diameter. Clemmer saw going cave; what lay beyond provided more encouragement after a few hours of work than the downstream dig had produced in days.

There was another shallow hole about 15 feet to the left of the one where Clemmer had been working. Stone put diggers to work in there, too. The first hole was dubbed the "right-hand dig" and the new one the "left-hand dig." Clemmer, toiling now in the left, squeezed through a narrow opening at the hole's far end. He wormed farther in and was digging out the floor to enlarge the passage when he spotted a mass of daddy longlegs hanging from the ceiling inches above his face. Seen from a distance, a daddy longlegs colony looked like a giant black beard growing on rock. Seen from inches, the insects looked like big, scary spiders. For the life of him Clemmer could not remember whether they had a venomous bite. As long as he did not breathe or move, the colony was quiet. When he did either, the black beard burst into frantic motion. It could have been a scene straight from a Stephen King novel.

While trying to lie absolutely motionless, wondering what a thousand daddy longlegs could do to the unprotected human face, Clemmer felt wind. It was a strong, steady breeze, as though he were lying in front of a house fan, and it was the closest a cave came to waving a flag proclaiming, ENTER HERE. It was such an exciting discovery that he forgot the daddy longlegs and scooted right out to tell the others. Everyone, Bill Stone included, agreed that this was damn good news indeed.

Despite their proximity, the two descending tunnels presented different challenges. Diggers working the right-hand tunnel were hacking and scooping their way through a tight crack plugged with mud, sand, and solid rock. It was excruciatingly difficult digging, but there was little danger of collapse. That wasn't the case in the left-hand dig, where, despite the winchers' best efforts, much of the ceiling still consisted of delicately balanced boulders that could dislodge at any time.

The explorers had to be steel-nerved to keep bashing around in there, hour after hour, day after day. Andi Hunter spent one such long day kneeling in the tunnel with just inches of clearance between the top of her helmet and two massive boulders. The giant rocks leaned against each other like a couple of shoulder-to-shoulder skid row drunks, each propping the other up and neither secure in its own footing. The seam where the boulders leaned together was directly over Hunter's head. If one or both came down, she would be squashed like an ant under a boot heel. Working with surgical caution all day, she managed to avoid that fate, but it was even more exhausting and stressful than usual.

By the end of Saturday, February 21, Clemmer and his mates had dug a shaft 15 feet deep and about the diameter of a big truck tire. There was good news and better news: no daddy longlegs and strong, steady airflow blowing out of the hole. Clemmer felt that airflow was one of caving's golden clues. There was even a whisper of breath in the previously dead-calm right dig. Things were looking up. Or, rather, down, and in the topsy-turvy world of supercaves, that was a very good thing indeed. Their sense of good fortune, however, would be short-lived.

FORTY-THREE

ON FEBRUARY 23, ANDI HUNTER WROTE in the expedition log: "The reports below chronicle what happened on the 21, 22, and 23 of February and give a sense of hope and despair that accompany every expedition undertaking real exploration."

That morning, Gregg Clemmer's group kept chopping away in the left-hand dig while others were burrowing into the right-hand one. Stone, with Andi Hunter and four men, left camp at dawn, heading for a new pit several miles to the north that had been reported the previous day by a local resident. After a few hours of exploring, they found the pit and rappelled into it. Only 300 feet deep, it went nowhere, but it was not without some exciting features. It contained a significant population of hand-sized spiders and giant millipedes, bright yellow and blue (coincidentally, the colors of the Ukrainian flag) and four inches long. When touched, they secreted foul-smelling acid from glands along the sides of their bodies. Jules Verne would have been pleased by such exotica.

Sunday, February 22, was market day throughout the region, so Stone and Hunter went shopping for seventy-five pounds of fresh food. When they returned, Gregg Clemmer was waiting.

"You want to see something neat, just wriggle on down that old left hole there," Clemmer urged.

By this time, the dig channel had veered under the riverbank, where trees grew. When Stone reached the tunnel's end, he found wind blowing out strongly enough to make the dangling tree roots wave around like giant insect feelers. *That,* Stone agreed, was neat indeed, even neater than Clemmer knew just then. Cheve Cave, far above them, breathed *in* all day long. This lower cave blew *out.* The air Cheve was sucking in above might be blowing out right here. And that might indicate the very thing they had come all this way to find: *connection.*

The next day, Stone decided to abandon the Star Gorge Cave. The digging team's Herculean efforts had added some length, but had found no moving air. Worse, they had begun to hit saturated sand, which meant, before long, a sump. That, plus the absence of airflow, convinced Stone to concentrate their forces on the two digs upstream at the sinkhole.

Even with extra hands, it was grueling work. The cavers contorted themselves into bizarre positions, then hammered and picked away, not infrequently standing on their heads, facedown in holes that, despite Hogan's oxygen pumps' best efforts, soon became stagnant. It was Charco all over again, and if it did not change soon for the better, it could become a spirit killer.

The work was already taking a toll, especially on the less experienced. One afternoon in late February, first-timer Ryan Tietz sat beside the entrance to the right-hand dig. His surroundings were grim: ugly chewed earth, lacerated roots, dirty light. A muddy five-gallon bucket sat in front of him. Coils of rope and power cords snaked around his feet. Over his head, the rocks were dyed bright green by some weird moss or lichen. Smeared with mud, he sat, heels drawn up, arms crossed, and elbows on knees, his face rough with two weeks' growth of beard, his eyes vacant and unfocused in what combat veterans call the "thousand-yard stare." Two weeks earlier, he had been bright-eyed, grinning, bursting with youthful energy. He was not grinning now, not by a long shot.

It was a measure of this kind of caving's severity that Tietz could appear so discouraged despite the fact that on the previous day, February 23, the happy digger John Kerr had made another encouraging find in the left-hand dig. Working at the tunnel's far end, he was chopping out compacted earth and pulling rocks free one at a time, as though dismantling a giant puzzle. He had been doing this for so long that the monotony had lulled him into a kind of trance. Then, finally, he pulled a big rock out of its socket and, lo and behold, there was neither soil nor rock beyond. There was nothing, in fact, but beautiful, beckoning empty space.

Playing his light through the gap, Kerr quickly determined that he had opened a hole in the ceiling of a bedroom-sized chamber below. Eureka! That was very good, but there was something even better: he felt a strong, cool wind blowing out of the hole, against his face. That meant a big cave beyond.

Kerr, Stone, and Hunter pursued the lead. They widened Kerr's hole and dropped down into the chamber below. At its far end, the passage continued for about 50 feet before widening into another chamber roughly 15 feet square and tall enough for them to stand in. From there, a brief descent took them to a 30-foot horizontal passage low enough that they had to crawl. Before long, they found themselves at the edge of a pit 75 feet deep and 10 feet in diameter. On the pit's far side, the horizontal passage continued; it would have to be explored as well. To reach that far side, they would have to rappel to the pit's bottom, then climb back up the far side. That would require rope, drills, bolts, and technical gear.

It was a major find that called for a meeting over lunch on the surface, where the news energized the entire team. Everyone in camp knew that the next day would be, in many ways, like a battle. It might end in disappointment and misery (and even death or injury, at worst) or it could be the "back door" to miles of passage that just might connect with Cheve.

FORTY-FOUR

THE NEXT MORNING, THE EXCITEMENT WAS palpable. People walked around camp with new spring in their steps and smiles on their faces. Today would tell the tale, whether long or short; it would bring some kind of resolution. Pauline Barendse led a team of three men down into the pit. At its bottom they explored a number of narrow, shoulder-width passages that descended like household stairs. One led to a hole that was blocked with rubble but could be dug out the next day. On their way back, they climbed the pit's far side and located a window high enough to require a bolt climb. If they had not yet unlocked Cheve's back door, they at least had not dead-ended.

Using computerized projections based on the team's survey data, Stone determined that the right- and left-hand dig passages would eventually merge. Not soon, but it would happen if they just kept digging. Unfortunately, work on the right had produced no rewards like those of the left-hand dig. It was Sisyphean labor. The point person, often Andi Hunter, filled a bucket with gluey mud and yelled. Team members behind her hauled it out. Before it was out of sight, another bucket was pushed down to Hunter, who filled that one and sent it up, then did the whole thing all over again. When Hunter tired, another digger took her place. As the pit deepened, foul air became an increasingly serious problem because there was no wind in this dig. They attached a longer tube to the oxygen generator and kept on working.

On February 26, over on the left, after some hammer-and-chisel work, Pauline Barendse squeezed through a hole that unlocked almost 150 feet of new passage—which, as passages in caves are wont to do, then just stopped.

On top of everything else, it began to rain. Water flowed down into both digs but was most troublesome in the right, where it streamed down into the hole occupied by the upside-down point digger, flushing mud into ears, eyes, and mouth and adding another layer of misery. It kept raining. And raining. Everything that goes up must sooner or later come down, including team spirit. To Bill Stone, it was obvious that three days of steady rain, on top of the unrelenting labor down under, was killing morale. One of the biggest encouragements had been that strong wind blowing out of the left-hand dig. But it had gradually weakened, then simply stopped. That may have had something to do with a pressure change occasioned by the weather system that had brought rain. Whatever its cause, to have such a strong, steady breeze just disappear was unusual and unusually discouraging.

Without the wind to follow, Stone knew, their work underground was now more educated guessing than logical pursuit. The fact was not lost on other expedition members. Even the perennially sunny Andi Hunter was hurting. After the second day of steady rain, she just holed up in her tent and skipped dinner. If Hunter was down that low, the others' spirits had to be in worse shape.

On the third morning of rain, Stone wanted to investigate the dome of the pit that Barendse had found in the left dig. He took along Hunter and Jim Brown, "Inspector Gadget," so called because of his fondness for, well, gadgets. He might have seemed an odd choice—while the Inspector was a world-class cave diver, he was much less experienced using a rappel rack, mechanical ascenders, and a heavy drill. But Stone had an ulterior motive. As the years and expeditions had passed, he had become better at monitoring his team members' physical and psychological states. On one expedition, he had seen the partly deaf, introverted Brown become completely alienated from everyone else. That could be like putting a spoonful of vinegar into a bottle of fine wine; the vinegar did not sweeten, but the wine was invariably fouled.

So it was with groups, and he could not let that happen here. He wanted to help Brown by spending some special time with him.

They set off for some climbing. Hunter was an experienced mountaineer and rock climber, accustomed to dicey leads on high faces. Clambering with big packs of rope and vertical gear through the left-hand dig, where so many boulders appeared to be held in place by nothing more substantial than sand and gravel, was nerve-racking. Stone, whose experienced eye gave him an edge here, estimated that it would take just two hammer whacks to dislodge some of the huge rocks and precipitate a cave-in.

The trio made it past the boulder gauntlet uncrushed, but at that point Stone had to return to base camp on an errand. That left only Hunter and Brown to do the climb, and that really meant Hunter. Even with her strength and skill, it would be challenging. Hunter put her climbing harness on over her mud-slimed caving suit. Then she slipped a sling over one shoulder and hung from it the hammer drill, battery, blow tube, wrench, hangers, bolts, screws, dynamic climbing rope, and static caving rope. All told, the equipage came to more than fifty pounds, heavy enough for a fit backpacker on the surface, a brutal load for vertical work in a cave. Bolt climbing was rather like setting rebelays in reverse, going up instead of down. To start, she drilled a hole as far over her head as she could reach with the heavy drill-battery combo, placed the hanger and bolt, clipped two étriers to the hanger, and stepped up into the étriers' foot loops. Then she did it all over again and again, and again. Each new bolt took her higher off the pit floor. It was an agonizing, and agonizingly slow, way to gain height.

Hunter had bolted her way to within two placements of the dome's top, a truly valiant effort, when she ran out of gas. Fortuitously, at that moment she spotted a small ledge, hooked her butt onto it, and finished her bolt work that way. As if on cue, Bill Stone returned down below. He climbed the rope Hunter dropped down, and together they surveyed a number of tight passages—"cracks" would be a better description—emanating from the top of the dome. Bolt climbing is the vertical equivalent of digging, dangerous and exhausting work with no guarantee of reward at its end, and they reaped no rewards here, despite all her hard work. She, Stone, Brown, and the other teams did not get back to camp until ten o'clock that night. When they did drag in, they were exhausted and discouraged.

By February 29, with no dramatic discoveries in Star Gorge, Stone decided to start hedging his bets. He, Hunter, and John Kerr began scouring the surrounding high country. It was decidedly unfriendly terrain. Time and water had carved the region's limestone surface into a dangerous jumble of holes, ledges, and stalagmite-like spikes. Between the ledges and the ridges were depressions of varying depth, with cacti and sharp rocks at the bottoms of most—natural booby traps. It was also snake country. Overall, it was a bad place to explore and a worse one to fall in.

They were not wandering around randomly with their fingers crossed. From education and experience, speleologists and geologists are able to "see" beneath the surface of the earth, the way we might see ocean bottom through very clear water. Stone and the other two hacked their way up a mountainside of hard, metamorphic rock until they found a strip of limestone. The meeting of these two types of rock creates a kind of "golden zone" where caves form. The impermeable metamorphic rock, which water cannot penetrate, channels flowing water to the softer, soluble limestone, where, sooner or later, it finds a crack or hole. If a variety of other geological stars align, it creates a cave. Because streams and rivers flow more or less continually, scientists who search cave country with limestone substrata like these look first to the agents of change, those rivers and streams, and follow them in the hope that they will disappear into the ground. Before long, Stone's team found a three-foot hole into which part of a stream flowed and disappeared. Like most other leads, however, this one proved false. Wherever that water went, it was no place humans could follow.

They pressed on like this for several more hours without finding any going caves, but at one point their search did produce a revealing moment for Stone. As Andi Hunter, in the lead, was hacking away through thick brush, Stone saw printed in white on the back of her blue T-shirt a map of Cheve Cave. Members of the previous year's expedition had all been given the shirts. At that moment, with his intimate knowledge of Cheve Cave's relation to the area's surface features, Stone realized that they must be standing directly above that cave's deepest point. Teased by Cheve, he could only shake his head and take out his frustration on the undergrowth.

Their reconnaissance having failed to produce anything of value, Stone began considering another sinkhole as a last resort. This one, which he knew offered going cave, was in the bed of the Aguacate River about a half mile southwest and uphill of the village of San Francisco Chapulapa. A recon team that Stone had accompanied had first found the Aguacate River Sink in 1989. In 1994, an offshoot group from the main Huautla effort penetrated it a half mile and almost 600 feet deep before a sump stopped them. Though near the nightmarish Charco Cave, the Aguacate River Sink Cave was more user-friendly, 80 feet wide and 50 feet from cave floor to ceiling. The fact that it aligned with the same heading as Cheve Cave, 330 degrees, led Stone to believe that it was another "stitch" in the long but still disconnected line running all the way down to the Cheve resurgence.

The days had stretched into weeks, and the team had been working itself into exhaustion. Bill Stone wanted to avoid another expedition-ending mutiny. It was time, he decided, to call a meeting.

FORTY-FIVE

ON MARCH 3, STONE CONVENED THE entire team to discuss the expedition's next move. He complimented everyone on their incredible commitment and hard work, reiterated their accomplishments, such as they were, and laid before them the available options for continuing. With little progress being made elsewhere, the group elected to focus on the Aguacate River sinkhole and the high, waterless "death karst" region (as they had started calling the unfriendly area) Stone and Hunter had braved earlier. Though the hike had been difficult, it had revealed thirty-five new pit entrances. On Saturday, March 6, the entire crew moved four miles southwest to San Francisco Chapulapa, a rough little village that made Stone think of a Clint Eastwood spaghetti western, complete with dusty streets, stray dogs, and ramshackle buildings.

Sunday, March 7, began the expedition's fourth week. John Kerr, Ryan Tietz, and a group of Poles who had arrived a few days earlier took one of the team's two-way radios and climbed up into death karst country to establish a high camp. Meanwhile, Bill Stone, Andi Hunter, and Jim Brown went into the Aguacate River Sink Cave. To do so they had to bypass yet another cave danger. Coral snakes, their red-and-yellow beauty belying the fact that they possessed one of the most lethal neurotoxins on earth, had been seen in this cave.

Luck was with them that day, because no snakes were waiting to bite unwary feet touching down on the cave floor. After the big entrance shaft, the roof dropped quickly, leaving a crawl space through which the Aguacate River flowed into this cave. Thanks to the incessant rain of the last few days, every stream and waterfall was swollen and roaring, but there was still enough airspace between the river's surface and the low chamber's ceiling to allow passage. As it turned out, a short crawl brought them to a huge waterfall plunging over the brink of the first of four shafts that would take them ever deeper into the main cave.

The plunging water in all of these drops was impressive, and Andi Hunter's rappel of the second pitch produced what may be the single most memorable image of the expedition. James Brown, hanging to one side, photographed Hunter in profile. Her yellow waterproof suit and red helmet glowed brightly against the cave's dark wall. Solid torrents of water beat down on her as she rappelled. Her eyes were squeezed shut and her mouth was flung wide open, as though caught in mid-scream. All around her droplets of water caught the light from Brown's flash, glittering like tiny suspended rubies, and a huge boulder in the foreground, polished by millennia of falling water, glistened like polished agate.

Their last rappel deposited them in the cave's main tunnel, half a mile long, 30 feet wide, and 30 feet high. They soon arrived at another powerful waterfall, this one pouring down from a cylindrical chamber that rose, like the bore of an immense piston, into the darkness above them. The flow here, Stone estimated, was twice as powerful as that coming through the entrance. There were only two possibilities for its source: either a sinkhole farther upstream or water from feeder caves higher on the mountainside. It begged for exploration, but they had no drill or bolts to climb up the chamber's wall. Walking on, they arrived at a sump, where Jim Brown, happy at finding a bit of *his* element, hopped right in to probe for submerged passages. Alas, he found none.

Dejected, they headed out, but before they had gone halfway Andi Hunter spotted something.

"Hold on," she said. "Check *that* out."

It was a slot, about half as wide as a computer monitor, where the cave floor and wall met.

"Give me a minute," she said and promptly slithered through.

"You're not going to believe this," she called back. Stone followed, forcing his big, rangy body through with some difficulty, and soon they were together at the upper edge of a sizable room with steeply sloping walls composed, in places, of hard mud. Cutting steps like mountaineers of yore, they climbed down to the chamber's bottom, where a hole 3 feet across extended down into darkness beyond the reach of their lights. Here, again, was reason for hope. Without ropes, they could not enter it, but they would come back.

That night in base camp, Stone was encouraged. The new chamber, called Andi's Room, aligned with the general heading, 330 degrees, of all the other stitches that Stone believed would ultimately connect and lead down to the resurgence.

"If I'm reading this map right," Stone said to Hunter, "I think I know what might come next. The passage will jog east, descend a little way, then pick up the 330 heading again—*beyond* the sump."

He did not need to add that if they bypassed the sump, they might find the missing link to Cheve Cave itself. They had to investigate, but first they would need to establish a camp down in the cave. They bedded down on the surface, feeling more optimistic but also more anxious. It had come down to this: the chamber beyond Andi's Room was their last hope.

FORTY-SIX

STONE AND HUNTER WERE JOINED THE next morning by Bart Hogan. They all packed for what they hoped would be an extended stay in the cave. Jim Brown and a friend of Stone's, the Mexican caver José Antonio Soriano, would act as Sherpas in support of the other three, packing supplies to underground camps as needed.

At around noon, Stone and his group descended into the Aguacate cave. During her initial rappel, Hunter experienced what every female (and long-haired male) dreaded, getting her hair caught up in the rappel rack. There were only two means of escape: cut the hair or pull it out. The stoic Hunter did some of both. They established Camp 1 at a depth of 560 feet.

For three days they worked at the bottom of Andi's Room, digging, moving some rocks, breaking up others. On the fourth day, John Kerr showed up with Jim Brown, José Soriano, and good news. The death karst team, exploring up in the high country of cacti and pits, had found some impressive caves, pushing one down to almost 400 feet.

Stone was relieved. He could hear the Big Clock ticking. They had just three weeks left to make some kind of breakthrough. And then, on March 15, time nearly ran out for him—literally.

Stone, Hunter, Tietz, and Brown had backtracked in the cave to investigate a high dome upstream. Stone thought it a good time for the other two men to refine their aid-climbing skills. Tietz went first and performed flawlessly. Then it was Jim Brown's turn. He climbed up using the bolts Tietz had placed and stopped to rest about 50 feet up. Something happened—afterward he was not sure what. But the big drill, with its dagger-pointed twelve-inch bit, came loose. Andi Hunter watched, horrified, as the drill dropped, narrowly missing Stone's head. He was wearing a helmet, but the drill would have cracked both it and his skull like eggshells.

The next day, Tuesday, March 16, produced perhaps the best and the worst moments of the expedition, revealing how quickly both prospects for success and personal relationships can change in a cave. That morning, everyone feasted on pancakes that Hunter had cooked on top and secretly brought down, along with a big jug of real maple syrup, for a surprise relief from the freeze-dried routine. A true team player, Hunter was known for this sort of thing; she'd take on the worst, hardest work, and then do something more.

After breakfast, Hogan and Kerr went down to the farthest point reached on the previous day to dig for new leads. Brown, Stone, and Hunter stayed behind to do the less rewarding but necessary surveying work. They soon received an unannounced visit from a couple, well-known veteran cavers both. After explaining the expedition's work thus far, Stone suggested that they could be of good use at the point where Hogan and Kerr had been laboriously working for hours and would undoubtedly welcome some relief. After giving it some thought, the two newcomers started down, taking Soriano with them. It was not until 4:00 P.M. that Stone and Hunter's survey work brought them, as well, to the location where Hogan and Kerr had been digging all day. There followed a very peculiar exchange.

Hogan and Kerr were standing by a nasty-looking crawl space, no more than 18 inches high, that they had opened. Assuming they had dead-ended on the other side, Stone said, "I'll throw the book through and you guys shoot the last [survey] shot."

Kerr and Hogan exchanged glances, then started giggling like little kids. Kerr said, "You have to come through and see this," after which he and

Hogan slid right through the little opening down by the floor. Hunter went, too; then Stone, scraping and scrabbling, finally got through himself. The others were just standing around, grinning. Clearly they were messing with him, a joke on *El Jefe*. They were all jammed into a space no bigger than a powder room, with no obvious exit. Stone was tired and scraped up and impatient.

"All right, good joke, let's wrap this survey and get back to the dig," he said.

"That's going to take a while," Kerr replied. "You've got a lot to survey."

"A least a kilometer," Hogan put in.

"All right, all right, stop bullshitting me," Stone said, trying to keep the fatigue and irritation out of his voice. It had been a long, grinding, unrewarding expedition. The joke was falling flat. But Kerr was grinning hugely.

"No shit, we busted this sucker wide open. We're past the sump and are down to a new stream with maybe half the flow of what we have at camp."

Then Kerr extended his hand and shook Stone's, who suddenly realized this was for real. "I don't bullshit about something like this," Kerr said. "Congratulations, man."

It suddenly hit Stone like a dropping boulder. *Breakthrough.*

Hogan and Kerr explained then. It turned out that there was an exit from the powder room. Earlier, they had been able to walk and crawl more than half a mile beyond the squeeze Stone had just come through. *Scooping booty every step of the way.* Having gone that far, they realized that theirs was a major find, and thus required surveying. They had stopped at the edge of a stream before returning to the place where they all stood just now. Soriano and the other two newly arrived cavers had expressed a desire to see the new passage and were off doing that just then.

Before going on, they sorted things out. Kerr would guide the surveyors, Hunter and Stone, forward through the convoluted cave passages. Hogan would stay behind and enlarge the vise-tight crawl space.

Andi Hunter was as energized as the rest, but before long she noted a conspicuous absence. *Where are those other three?* According to Hogan and Kerr, they were going "only a little way downstream." But they had been gone several hours. Hunter feared that they were pressing on into virgin territory, "scooping booty." *That* privilege belonged to the people who had worked so hard opening the way to it. It was disconcerting, a violation of the unwritten code. Bart Hogan, veteran of the 2003 and '04 expeditions, as well as many others, put it like this: "The worst ethic is to run in there, do no survey, and then run back out. Surveying is the most important thing."

Hunter voiced her concern to Stone, who shrugged it off, focused intently on the mission, as always. As the hours passed, she became increasingly disturbed, repeatedly noting their absence to Stone.

After four hours, shaking and in tears, Hunter confronted Stone. *Those others just got here and they are scooping what we have all worked so hard for!* Stone admitted that it was possible but said there was nothing to be done about it now. Hunter could only fume.

The absent trio reappeared at 7:00 P.M. They had done no surveying. Neither Hunter nor Stone spoke a word to them. Stone did take his friend Soriano aside and quietly ask, in Spanish, that he and the others, if they came back, take over the surveying the next day. Translation: "We don't want you in there running ahead of everybody again, amigo. *Get it*?" Soriano got it. The three went on their way. Shortly after, Hunter handed her tools to Stone, saying, "This isn't fun anymore," and went alone back to camp.

FORTY-SEVEN

TWO HOURS LATER, STONE ARRIVED IN Camp 1 himself and learned more. The couple had told Bart Hogan that they had not gone beyond the point where Hogan and Kerr themselves had stopped, the edge of a small stream. If that was true, they had not really explored virgin passage. They would have only retraced Hogan and Kerr's footsteps. Stone was glad to hear it; Hunter was not the only one upset.

At the end of that day, the newcomers wrote in the expedition log, "After reaching the end of today's exploration [meaning, presumably, the stream where Hogan and Kerr had turned around], we reluctantly returned to cave camp, then on to the surface." In other words, they had not passed into virgin territory.

On March 18, Stone and the others descended to that stream and beyond. Later, he noted in the activity log, "it was obvious from the boot prints that they had gone much further." There was nothing anyone could do, but no one was happy about it. John Kerr said, "You know it's bad when you have to rank your friends' character by the number of meters that constitutes 'a little way downstream.' "

STONE'S TEAM PUSHED ON TO A chamber a full mile beyond the cave's "terminal" sump. The tunnel soon constricted, forcing them to crawl on their bellies. This was John Kerr's element. He took the lead with his titanium tool and burrowed 150 feet forward. Hogan, crawling a good distance behind, called to Kerr. No reply. Hogan called again and again, with no response. He was beginning to fear the worst—that Kerr might have suffocated—when the other man finally came backing out with frightening news. The tunnel's ceiling had collapsed on him, burying the length of his body to eye level.

"I had to back out fast to get a breath," he reported calmly. Nevertheless, he had just had as close a call as Stone's recent brush with a falling drill. If the ceiling collapse had buried his arms and face—which it had come close to doing—Kerr would have died a most unpleasant death by premature burial.

There's no indication that the cavers viewed Kerr's experience as a bad omen, but subsequent events must have left some of them wondering. During a trip to the surface from Camp 1, Andi Hunter planned another culinary surprise for her teammates. She cooked up a big pot of black beans and brought it back down to camp, where she added spices, sauce, and sausage for a chili con carne feast.

Everyone ate heartily, then bedded down for the night. Before long, little depth charges started detonating in Hunter's stomach, and she was not alone. Soon the camp was resonating with thunderous flatulence, and then sufferers started rushing to and from the latrine, 100 feet away, in a steady stream. Before long, violent vomiting added itself to the vicious cramps and diarrhea. Food poisoning had sabotaged Hunter's good intentions.

Stone was singled out for particularly devilish punishment. While Hunter was off at the latrine, Stone used his pee bottle. Conserving light, as always, he missed the bottle and urinated instead on their sleeping bags. It was not his first unpleasant encounter with pee bottles. Two days earlier, again in the dark, he had picked his up, taken a long, lovely piss, screwed the bottle's cap on, and gone back to sleep. The next day, foraging for some granola bars, he'd opened their Nalgene bottle container and found them marinating in vile yellow liquid.

When Stone himself finally made it to the latrine, he found it overflowing with vomit, runny excreta, and toilet paper. He also found that they were down to their last few sheets of the latter. Wet with his own urine, his gut clenched, barely suppressing the urge to blow vile matter out of both ends, he and Kerr dug a new hole and put it to use immediately. So much for the romance of exploration.

ON MONDAY, MARCH 22, still without a major breakthrough, Stone, Hunter, Brown, Kerr, and Hogan returned to the surface after more than a week underground. Disappointed that their own effort had produced nothing, they were happy—"overjoyed" might be a better term—to hear that the death karst teams had explored one of their caves to more than 1,000 feet deep. Two cavers, Pavo Skoworodko and Artur Nowak, remained up there, but they had no more rope.

On Wednesday, the five cavers, followed by a burro train, ascended to the high death karst camp, at 8,000 feet. They passed through strange terrain, forests of hundred-foot trees festooned with green moss and snakelike vines, and through thickets of a plant called *mala mujer* (evil woman). It was a pretty thing, with delicate white flowers and shiny, maplelike leaves, but poisonous spikes on its stems and leaves can inflict the worst sting of any nettle-type plant known, causing extreme pain and ugly rashes that can last for days, leaving ugly blotches that often remain for months.

The high camp occupants had named the new cave J2. *Jaskinia* is the Polish word for cave. J2 was still going. Stone well understood that this was their endgame. Cheve was blocked at its deepest point by breakdown. Charco ended in a terminal sump. The Aguacate River Sink Cave had tightened to the tunnel that almost killed John Kerr. The Star Gorge caves had gone nowhere. This was their last, best hope. They desperately needed J2 to go.

Five days later, a final-push team spent eighteen hours in J2, reaching a maximum depth of about 1,500 feet. That was good news, and the cave was enlarging the deeper they went. The final passage was about 15 feet wide and 50 feet high. At about 1,300 feet deep, several streams joined together to produce a powerful river that they followed to the edge of a huge chamber. There, the river shot out into black void. A strong wind blew down into it, as well. The cave clearly went much, much deeper. But, out of rope, they could go no farther.

Once again, many weeks of excruciating labor and extraordinary risks had produced more frustration than fruition. At the bottom of J2 were encouraging signs—growing passages, increasing water flow, and wind. They had to be going *some*where. Since J2 was between the main Cheve Cave and its resurgence down at the river, it was reasonable to expect that it might turn out to be the missing link, or a part of it, that Stone had been seeking.

But it had not done so, and another twenty years might elapse before it did. Spirit unbroken, Stone—astonishingly—vowed to return, saying, "We have no choice but to go back and leave J2 as a target for a follow-up expedition." His determination was admirable, but despite decades of work, he had not proven Cheve to be the deepest cave on earth.

It remained to be seen, halfway around the world, what Alexander Klimchouk and his cavers would prove about Krubera in 2004.

FORTY-EIGHT

ALEXANDER KABANIKHIN'S ACCIDENT HALTED THE EXPEDITION of August 2003. Alexander Klimchouk planned to lead one of his own—the sixth since 1999—to Krubera in the summer of 2004. In August, his team arrived on the Arabika Massif with fifty-six cavers (forty-five men and eleven women) from seven countries, ten thousand pounds of supplies, and two miles of rope.

They, like all those who had gone into Krubera before them, had no illusions about their coming weeks. Cheve Cave was full of challenges and hazards, and there was no hyperbole in the assertion that exploring it was like climbing Mount Everest in reverse. But neither was it uncommon for people to wax poetic about the exquisite turquoise pools of the Swim Gym, the Turbines' polished, gleaming walls, the daunting splendor of Nightmare Falls. The *beauty* of Cheve, in other words.

Not so Krubera. Aesthetically and technically, it was an ugly cave, tight, wet, freezing, and unrelentingly vertical. Cheve presented cavers with a modicum of walking passage and nontechnical descents like the A.S. Borehole and Low Rider Parkway. In contrast, about 90 percent of Krubera was technical (requiring ropes and hardware), one long elevator shaft after another pounded by glacial-melt waterfalls, connected by tortuous meanders. Cheve gave its explorers room to back off and take in the vistas, like tourists on the Blue Ridge or the rim of the Grand Canyon. Not Krubera, which almost never relaxed its tight embrace.

Be that as it may, these explorers encountered beauty of another kind: the elegance of superb leadership and organization. People had clearly assigned tasks, understood what they were, and doing them well for the good of the team was a point of immense pride. Moreover, on Klimchouk's expeditions, cavers rarely got hurt. Women as well as men were present, but sexual shenanigans were conspicuously absent. Even couples slept separately, to avoid making others feel deprived or uncomfortable.

Klimchouk's August 2004 expedition began with typical quiet efficiency. Team members acclimated once again to the supercave environment, rigged all the drops, established and stocked all the camps, drilled and blasted wider passageways. They also completed installation of a telephone line to the deepest camps.

After three weeks of this work, they reached a sump at 5,823 feet. Gennadiy Samokhin, who was to eastern European cave diving as the great American diver Jim Brown was to American, geared up and dove the sump's 32-degree water. It was about 35 feet deep, and at the bottom he found a tight slot. Water went through it, but he could not. That left two options: work underwater with hand tools in zero visibility to widen the squeeze or find a way around the sump.

Sergio García-Dils, back for another Krubera close encounter, rappelled under a hammering blast of frigid water to the bottom of a narrow, nearby chamber, hoping it would go. No luck. Then two other expedition members, Dmitry Fedotov and Denis Kurta, found a tubular passage about 30 inches in diameter that descended steeply from one of the sections just beyond their deepest camp, some 415 feet above the terminal sump. If other places in Krubera were ugly, this hellhole was horrific. For more than 100 yards, there was not even room to creep on one's hands and knees. Over the eons, rushing water full of abrading gravel had cut and gouged the tube's walls, leaving knife-edged ridges around its entire circumference. There were also many rock spikes poking out at all angles. It was like slithering on one's belly through a twisting tube full of blades and daggers.

Fortunately, the cavers were amply rewarded. The passage (they named it Way to the Dream) bypassed the sump and kept going until it reached another at 5,888 feet. At the end of August, Samokhin free-dove this sump (meaning it was a breath-hold dive, without scuba gear), disappearing into the cold, opaque water. He did not resurface immediately, which was cause for both hope and fear. If he had popped right back up, it would have meant that there was no way past the sump. That he lingered longer meant that either he had found a way or was in trouble.

Samokhin's companions stood anxiously around the sump, their breath visible in the cold, damp chamber. After what seemed like a very long time, they saw a light approaching through the bluish water. Samokhin surfaced, breathless but grinning.

It goes, he told them. *There is an obstruction of boulders, but an opening that can be enlarged.*

Samokhin had just completed the deepest dive—free or with gear—ever done in a cave and had pushed Krubera's limit to 6,037 feet, establishing it firmly as the deepest cave on earth. This was a historic accomplishment. Samokhin and the others were well aware of that, and an ecstatic little celebration, with cheering and clapping and hugging, took place right there at the sump. When news of the find was telephoned up to Klimchouk and the others in camp, another celebration erupted.

Several days later, team members began crawling one by one out of Krubera's mouth. Klimchouk hugged each in turn, and others were waiting with colorful bouquets and mugs of good wine. With the whole team reunited, a third celebration ensued.

EVERYONE WAS OUT SAFE, AND GREAT new work had been done. All good. But, to use an Antarctic analogy, they had not reached the South Pole just yet—there was still more descending cave to explore.

They would not explore more of it on this trip, however. Klimchouk had planned for a month up on the Arabika Massif, and they had spent their month. Both supplies and cavers were nearing exhaustion. It was time to pull out.

FORTY-NINE

ONE HAPPY RESULT OF ALEXANDER KLIMCHOUK'S organizational skills was that his expeditions ran like Swiss watches. Another was his ability to mount multiple expeditions in a single year. There were at least four more weeks of acceptable caving weather in 2004, and four more iffy ones after that before winter really slammed down on Arabika. Klimchouk and his fellow explorers were determined to use them.

Klimchouk's August expedition paved the way for 2004's second Krubera effort. Between the August 2004 expedition and a previous one he had led in Turkey, Klimchouk had been in the field for two and a half months. He simply could not manage another month away from family, university, and laboratory. Someone else would have to lead the next effort, scheduled for October.

Yury Kasjan was the logical choice. He knew Krubera as well as anyone alive. Like Klimchouk, Kasjan was a fine leader, quiet but good-natured; had unbelievable stamina; and was technically adept at all aspects of extreme caving. Younger cavers liked him for his steadiness, his dry humor, his vast experience, and his concern for their safety.

Kasjan left home early on Thursday, September 30, 2004. On his way out of the house, he found a note from his son:

> Dad,
> I bet a pack of Snickers that you will find the way in the cave and make it deeper. So give it your best shot!

I will do my best, dear son, Kasjan thought to himself. But it would be no cakewalk. He had been to the Arabika ten times and in Krubera six. He had suffered its killing cold, shivered in its freezing water, inched along its endless crawlways, and rappelled its gigantic shafts. He had spent not just days or weeks but *months* down in its deepest, darkest reaches. He knew that this could well be the hardest expedition of all, not just in Krubera but also of his life.

Despite his years of experience, as Kasjan drove to the sparkling new steel-and-glass Kiev train station, he found himself worrying more than usual. Down here at sea level it was still late summer, but they would be spending October at 8,000 feet in the Caucasus Mountains. Winter conditions would definitely begin in November. October was a toss-up—mountain weather was notoriously fickle. Would the serious snow and windstorms hold off long enough for them to complete their expedition? Even if it didn't snow in October, it would rain, exacerbating the risk of dangerous flooding. And what would they find beyond the sump that had stopped Gennadiy Samokhin in August? Based on dye-trace experiments, they knew Krubera could go to 8,000 feet or even deeper. But could they unlock the route that *proved* this cave, once and for all, to be the deepest on earth?

Kasjan was also concerned about his nine-person team. By supercave expeditionary standards, this was a radically small group, the caving equivalent of a light, fast, alpine-style mountaineering attempt. (Ironically, it was a mirror image of the small team Stone had taken to Cheve earlier that same year.) All of the team members were fit, well-equipped, experienced cave explorers, but only Kasjan had been in Krubera before. How would the others fare in the deepest, most dangerous cave in the world?

One of Kasjan's young team members, Emil Vash, twenty-two, was wondering the same thing. The tall, slender Vash, who resembled the American actor Edward Norton, was pursuing a degree in physics at Uzhgorod National

University in far western Ukraine. Like Klimchouk and so many others, Vash had started caving early, at fifteen, with a Young Pioneer Palace. He met Yury Kasjan in 2000 and soon joined his expeditions. By 2001 he was leading his own groups into serious caves.

It was raining in Uzhgorod on Thursday when Vash boarded the train to Kiev, where he would join Kasjan and some of the others. He shared his compartment with a quiet man named Vasil, a talkative man named Sergey, and a pretty girl named Svetlana, who regaled the others with tales of her psychotherapy sessions. Like all long journeys, this one began with small adventures, a "wonderful atmosphere of understanding between unknown but not strange people," Vash wrote in his diary. As a newcomer to Krubera, Vash saw things with fresh eyes; this made his observations especially valuable. In addition, he was a gifted writer and his diary is the best record of the October 2004 expedition.

In Kiev, Vash disembarked and went shopping at Atlantida, a Ukrainian outdoor-gear emporium, to buy new caving coveralls, batteries, an ascender, and other equipage. He then met Kasjan and more cavers and they all took another train to the city of Sochi, the jumping-off point for Ukrainian Krubera expeditions. On October 3, under a cloudy sky spitting rain, Kasjan, Vash, and the other team members crossed the border from Russia into Georgia. Their passage was uneventful save for the scrutiny of one suspicious border guard, who thought that their dozens of small batteries looked like rifle cartridges.

Nine cavers would work underground in two separate groups. Team A included Kasjan, Vash, and three others. Blond Ekaterina "Katya" Medvedeva, twenty-three, was as strong and brave as she was pretty. Igor Ischenko, thirty-six, and Kyryl Gostev, twenty-one, rounded out the team. These five would be the expedition's spearhead, employing Klimchouk's "no dead ends" approach. Team B, which would support the other group's effort, included Dmitry Furnik, thirty-six; Ilja Lapa, twenty-one; and Sergey Baguckiy, forty-two, all from Yalta; and Vladimir Dyachenko, twenty-five, from Kharkov.

Kasjan's plan called for the use of at least four underground camps that had been created by the August expedition, at 700 meters (2,297 feet), 1,215 meters (3,986 feet), 1,400 meters (4,593 feet), and 1,640 meters (5,381 feet). The schedule called for a week to descend, a week exploring at depth, and another week ascending. This was less time than normally required for a major Krubera effort, but Kasjan expected that the rigging left by the summer expedition would speed their passage.

As usual, the team members took great pains in sorting and packaging their food "modules," each of which contained rations for five people for two days. Bill Stone's cavers, in line with their leader's view that weight considerations were more important than comfort on expeditions, tended to rely on light but unpalatable freeze-dried foods in their deepest camps. Klimchouk preferred to work with caves rather than "attack" them; he believed, as well, that happy, comfortable cavers could explore more effectively than those strung out by Spartan conditions. In addition to staples like rice and pasta, therefore, Klimchouk's teams stocked up on comfort foods: candy, cookies, cakes, sausages, cheese, canned vegetables, condensed milk, pâté, *nishtyak* (a gorplike blend of raisins, apricots, prunes, figs, and candies), and, of course, spirits for appropriate celebrations.

On October 4, the expedition established its base camp. Emil Vash and Vladimir Dyachenko, lacking tents, holed up in a small, icy cave. Vash's dinner that night included hot milk laced with butter and honey, to ease a hacking cough brought on by the altitude and the day's exertions. Snow covered the ground around their foxhole-like abode. Both men were anxious to get down into the cave, where conditions would be appreciably more comfortable.

FIFTY

THE NEXT DAY, VASH, KASJAN, MEDVEDEVA, and Ischenko entered Krubera. They rappelled pit after pit, struggled through the Crimea and Mozambique meanders, dropped their freight bags at about 1,200 feet deep, and then headed back up. The walls of the meanders had a bizarre, pockmarked look that reminded Vash of Abkhazian buildings, blasted and pitted by machine-gun fire and RPG rounds. Kabanikhin rescuers had drilled into these Krubera Cave walls to place explosives in order to blast open passages too tight for the cumbersome litter.

They were finding significant problems with the rigging—jammed knots, old rope, too much distance between belay points. These were remnants of the summer 2004 expedition, and not the usual leavings of a Klimchouk team. But that group had included almost sixty members of widely varying ages, experience levels, and national origins. That made the sloppy work more understandable, but no less frustrating. For Kasjan and his smaller team, it was like trying to untangle a giant fishing-reel backlash one maddening snarl at a time, except that they were doing this in the dark, hanging from sheer walls with hundreds of feet of empty space yawning beneath them, awash in nearly freezing water that numbed their fingers in minutes.

It all put the normally sunny Vash in a rare bad mood. "I was hanging for a long time in doubt, re-doing and securing again," he wrote in his diary later. "Finally we have finished at the old campsite at the depth of –500m." It was from here, the 500-Meter Camp, that Kabanikhin's rescuers had stabilized him and begun his long carry to salvation.

YURY KASJAN REMAINED IN THE CAVE late that day, after the others had retreated to the surface. He often worked alone, relishing the feeling of being one with the cave and grateful for the opportunity to work free of distractions. He was about 1,150 feet deep, free-climbing a 10-foot vertical pitch up to a ledge, when his foot slipped off a wet, slick nubbin and he fell, wrenching a knee. It was not an incapacitating injury, but it easily could have been. If nothing else, the mishap illustrated how every move in a cave, even the seemingly routine, could do you in. Emil Vash was cooking dinner up top when he and other team members heard about their leader's slip and fall. It was somewhat unnerving for these novitiates to learn that an icon like Kasjan could come to grief so early in this effort.

Though Vash was thus far uninjured, he was suffering as well. He had been told that ascending cavers could go up from the 500-Meter Camp with two sixty-pound bags, twice a day, and feel fine. Based on what he was experiencing—and Vash *was* a fit, veteran caver—he found that hard to imagine. Moreover, on his way up, Vash discovered how, in supercaves, fatigue and chance can combine to produce unnerving consequences. A few hundred feet above the 500-Meter Camp, he came to a rebelay station where he had to hang from a short fixed rope while detaching his ascenders from the lower climbing rope and reaffixing them to the one that ran on up beyond the rebelay. As he was hanging on the short rope, one of the carabiners fastened to the rebelay bolt in the wall twisted and opened, and out popped one of his safety ropes. Vash had taken the technically correct precaution of reattaching his ascender to the rope above the rebelay station, and that saved him. Had he been hanging only from the "rope rail," with its defective carabiner, he would have fallen to his death.

Vash made it to the surface without further incident. It was full dark, with bright stars shining overhead, when he emerged. He had a quiet dinner and mugs of sweet tea with Kasjan and Medvedeva. The telephone system picked up random signals from various radio stations, and Vash found the intermittent snatches of music and talk reassuring evidence that human civilization was still out there.

In company with various team members, Vash spent the next days carrying supplies and rigging ropes deeper into the cave. This was not simply a matter of repeated rappels. They had to haul heavy bags through horizontal meanders as well, those maddening tubes that required crawling on hands and knees, and sometimes on bellies. The aptly named Sinusoida Meander was representative. About 650 feet long, it descended 330 vertical feet in a series of cramped, narrow, winding passages that, on a map, look like the nasal passages of a giant with a badly deviated septum. After one such hard haul, Vash reached the 700-Meter Camp at 1:00 A.M. somewhat the worse for wear, with a bad cough, an aching back, and a wrenched knee.

Clearly, this was like no other cave Vash had ever entered. After what sounded suspiciously like a minor attack of The Rapture, he confided to his diary, "In general, I can't feel any special spiritual closeness with [Krubera] yet. That reminds me of some kind of absence of imagination; I look inside myself and then ideas appear about my own callousness and roughness."

Determined to persevere, Vash tried to relieve his anxieties by recalling the surface. He fell asleep at the 700-Meter Camp thinking of Arabika's lush greenswards and meadows so strewn with wildflowers they looked like one of Monet's impressionistic canvases.

On October 9, Team B ferried loads down to the 500-Meter Camp while Vash's lead team, starting at 11:00 A.M., began to work down to the 1,200-Meter Camp. It was a tricky route, involving vertical pitches washed by frigid cascades interspersed with more tight, twisting meanders. At eight o'clock that evening, Vash, Kasjan, and Medvedeva occupied the 1,200-Meter Camp, which Vash found surprisingly commodious—dry and roomy and equipped with a tent big enough to hold six comfortably. Vash made minor repairs to his vertical gear, snacked, and listened to the question that would not stop echoing through his brain: *Will I get to the bottom?* And listened, as well, to its inescapable evil twin: *Will I get back to the top?*

The camp, a Hilton by expeditionary caving standards, had a stocked pantry with one hundred person-days of food, or enough for the five-person Team A to remain underground twenty more days. That evening, dry and warm in his blue sleeping bag, Vash dreamed not of wildflowers and meadows but, oddly, of beautiful pictures in a grand museum. If there was some symbolism connected to his current adventure, he could not, upon waking, discern what it might be. But at least he was not having nightmares.

Day by day, like mountaineers climbing up to successively higher camps, Vash and his teammates descended to successively lower ones. On October 11 they occupied the 1,400-Meter Camp. From here, Klimchouk's summer teams had explored virgin cave all the way down to 6,037 feet (1,840 meters), establishing Krubera as the world's deepest cave. That lowest spot would become *this* expedition's jumping-off point. Everything previous was just preparation.

To reach that camp, however, they had to get themselves, their gear, and their supplies through a short but treacherous sump. The sump was just short enough that scuba equipment was not *absolutely* essential. But it *was* long and tight enough to stretch a breath-hold dive to its extreme limits. Because bringing scuba gear involved another whole dimension of logistics, Kasjan had elected to traverse this sump by breath-hold diving. Vash knew he would have to go through if he was to have any chance of reaching Krubera's bottom. It would be his first encounter with such a sump, and he would be attempting it deeper than he had ever been before, as well as in the most dangerous cave he had ever encountered—and quite possibly the most treacherous in all the world.

FIFTY-ONE

ON OCTOBER 12, THE DAY OF the planned sump crossing, Vash found himself too nervous to eat breakfast, an unusual experience for the hardworking and normally ravenous caver. At midmorning, hauling two bags, he started down, rappelling two long vertical drops, both flowing with strong and freezing waterfalls. During the descent, Vash's thoughts alternated between excitement, apprehension, and the stoic need to endure. Arriving finally at his personal Rubicon, Vash discovered that Kasjan had already made the dive and was on the other side. It was time for the other cavers to follow.

Gutsy Katya Medvedeva went first. She put on a diving mask, sucked in a huge lungful of air, and disappeared into the flooded tunnel. They waited several long minutes, but she did not return. Vash and the others assumed—*hoped*—that she was safely on the other side. Then, after what seemed to them a very long time, she reappeared, sent by Kasjan to assist the less experienced Vash and the others. She broke the surface with water streaming from her helmet, lights shining, and a reassuring smile brightening her lovely countenance. *Not so bad after all*, she reassured the others. *Just stay calm.*

Vash's turn came. He had been told that the underwater tunnel was about 30 inches high and 10 feet long. That didn't sound like much, but its third dimension was what made the tunnel so dangerous: it was just 18 inches wide, roughly the diameter of a large pizza. In addition, its walls were rough and spiked with protrusions that could easily snag a caver. It would not take long to drown in those conditions. He would do well to be able to hold his breath for sixty seconds in water that cold. Throw in the added factors of oxygen-devouring panic and wild attempts to extricate oneself, and the probable time before drowning dropped considerably.

Vash had a premonition that he would not make it through, but he pushed on regardless. He hyperventilated briefly to blow carbon dioxide out of his system, held a last breath, and dove. His premonition quickly came true. He didn't submerge deeply enough and, halfway through, his helmet caught on a ceiling obstruction. Panicking, he flailed back out of the sump and surfaced, unnerved and gasping. It had been one of the worst moments of his caving career. Though it had taken only twenty seconds or so for Vash to extricate himself, it had seemed much longer. Vash's exertions had consumed the oxygen in his system voraciously, causing him to feel suffocated much too quickly. Reflecting later, he could remember no specific thoughts during the incident, just a black space filled with terror.

Cavers are nothing if not persistent. Calming himself, Vash went in once again, taking care to go all the way down to the tunnel's floor this time. He cleared the ceiling but found, terrifyingly, that the sump was so narrow that it pressed against the surface of his body and helmet. This was a very tight scrabble while completely submerged in zero-visibility, 32-degree water. He had been told to expect such conditions, but as veterans of childbirth and combat can confirm, no matter how great the effort, thinking about such things is never adequate preparation for their reality.

When he finally popped out the other side, there was a weird similarity to birth about the whole thing: the painful passage through a constricted canal, followed by shocking emergence from water to air in a cold room where waited an oddly dressed man with lights on his head.

"Calmly, easy, it's all good now," Kasjan told the rattled Krubera rookie in a soothing voice. Vash was, of course, hugely relieved that he had survived; but he was almost equally dismayed that, in his role as resupply Sherpa, he would have to pass back and forth through this little chamber of horrors many times before all was said and done.

On the other side, Vash and Igor Ischenko helped Kasjan resupply the 1,400-Meter Camp, leaving additional food, batteries, fuel canisters for stoves, and other items. Before they were finished, Ischenko began to feel nauseated and weak and retreated to the next camp above. Vash and Kasjan finished their restocking, and then it was time to go back through the sump.

Kasjan went through first, leaving Vash alone on the other side, the deepest, and probably the loneliest, man on earth. Only one person had ever been deeper, and that was Gennadiy Samokhin, whose August 2004 dive had taken him to the bottom of the other sump, down at 5,664 feet.

With his heart racing and his fear under control—barely—Vash sucked in as big a breath as his chest would hold and dove in. It took longer than his first crossing had, and he surfaced on the other side trembling and panting. Once again, Kasjan, like a Ukrainian Marion Smith, was there to help.

"Quietly, quietly," he said with a reassuring smile and a pat on the shoulder. "All is okay."

Vash had been underground for days. He was soaked, freezing, and in absolute darkness held at bay only by his frail lights. He was not an experienced diver, and the chamber he'd had to pass through, while not long, was squeeze-tight and spiky and filled with opaque water. It had been the longest short period of Vash's young life, and would be so every time he made that passage.

Over the next few days, both teams brought more supplies to the deep camp, creating caving's version of the high camp in mountaineering, the final resting place before the launch of the summit assault. Working so hard that deep, Vash was discovering something as true in supercaves as it is for extreme mountaineers in the "death zone" of 8,000-meter peaks: the body does not really recover under such conditions but deteriorates more or less rapidly, depending on the individual. Most supercave expedition members lose a pound a day or more, and since few are fat to begin with, their bodies soon begin eating their own muscles. For Vash, the work was so exhausting that he found himself floating away at times, as had Andi Hunter in Cheve, his mind involuntarily detaching from the pains of cold, sharp rocks, burning muscles, and exhausted body. In such a fatigue-induced stupor, he often heard music and saw the faces of friends floating before him.

By October 14, they'd finished the preparatory work and had moved into the 1,400-Meter Camp, separated from the world by several miles of mostly vertical passages and the frightening sump. This was a far worse place to get hurt than the 500-Meter Camp, where Kabanikhin had been "lucky," if you could call it that, to suffer his own injury. Serious injury or illness this deep, beneath so many vertical sections, and on the far side of that infernal sump, meant almost certain death. Contemplating such reassuring facts, the team rested, ate, and drank. They were ready to begin the final, crucial stage of the expedition, a two-pronged attempt to find a way past the lower sump that had stopped Gennadiy Samokhin on his final summer dive. Two separate probes would explore passages that diverged at what they called the Large Fork, just above the Samokhin sump. One passage had already been surveyed, but not fully investigated, by the August group. The other was terra incognita.

Leader Kasjan further divided his A team. He and Medvedeva would press on into the new passage; Vash and two others would explore more carefully the passage surveyed by Klimchouk's August team. As had happened before, there might be windows or cracks that, if pushed, could lead onward.

Vash, despite his exhaustion, finally encountered beauty in Krubera dramatic enough to penetrate even his thick haze of fatigue. In this section of their exploration, he passed through gracefully winding meanders filled with limestone formations of fairy-tale beauty, delicately layered walls of white, brown, black, and red, alternating with sparkling waterfalls. The beauty had a beastly twist, though, because he and his already fatigued mates worked their way down to the terminal sump without discovering new leads. By the time they reached the sump, terminal exhaustion was setting in. Vash was finding it difficult to move. He was becoming hypothermic, his muscles stiffening like cold taffy, his mind sluggish. More than anything else, his body begged for sleep. It was a struggle to get back to the deep camp, where the team dined on hot tea, snacks, and macaroni and cheese, and collapsed gratefully into their sleeping bags.

FIFTY-TWO

THE NEXT DAY, TEAM B CAME down through the scary sump above and arrived in camp. With fresh troops, Vash and several others began the laborious job of surveying. Despite wearing dry suits, they had been exposed for many days to waterfalls and pools that had worked through the holes in their suits and the leaks in their ankle and wrist cuffs. Constantly wet, even with the warming respites in their sleeping bags, their bodies were slowly losing ground to the relentless chilling that produced a slow-motion, irreversible hypothermia. Before long, it trumped even Vash's need for sleep. Now, more than food or rest or anything else, his body was craving warmth, just as intensely as one parched by a desert craves water.

They had completed the transition from surface dwellers to troglodytes, cut loose completely from surface norms, including circadian rhythms. On October 15, they didn't start work until 2:30 P.M., finishing at about 9:00 P.M. Vash and most of the others spent the next day "slugging" in their sleeping bags, resting and trying to get warm. They were approaching the point of diminishing returns, which arrives sooner or later on all extreme expeditions, a time when the explorers feel like they are working harder and harder but achieving less and less. Vash captured this feeling of slow, ineluctable decline in words both disjointed by fatigue and yet strangely poetic, reminiscent of those penned by Robert Falcon Scott as he and his little party struggled across the polar wastes toward a fate they all knew could only be death:

> I was thick and tired from the process of the permanent delivering from superfluous moisture—those reactions of the organism for the cold conditions were very strain. But nothing could be done.

Time, supplies, and their stamina were all running low, which was bad enough. Worse, though, was the knowledge that, regardless of what they found, the hardest part of their ordeal was still to come: getting out.

While Vash and the others were recuperating on October 16, Kasjan and Medvedeva continued their exploration of the new passage beyond the Large Fork. The next day, the two, joined by Igor Ischenko, continued working in the new section, which they named Windows. In the beginning there were dry, downward-sloping passages, so tight Kasjan imagined they might have been made by large earthworms. These slanting meanders alternated with sizable pits (one was 175 feet deep) down to 6,048 feet, where they were stopped by a squeeze not even the svelte Medvedeva could pass through. Stymied, they started climbing back, but before long Kasjan spied an opening, on the far side of a shallow pit, that he thought might go. They rappelled down into the pit, climbed up its other side, and entered the new passage. It *did* go; they explored it down to 6,272 feet and, with the passage still extending, decided to return to camp and come back the next day.

Vash and others with him, tapping their last reservoirs of energy, had spent twenty hours surveying and exploring in the other passage leading from the Large Fork before returning to camp. They were happy to hear the good news from Kasjan and Medvedeva. The team celebrated by drinking toasts of cognac with lemon slices. But even with the bracing spirits, they were so tired, dirty, and cold that it felt to Vash like they were "in the Dragon's asshole." They now had food and fuel for just two more days, and energy that *might* last that long. Any breakthrough would have to happen soon. They were stretching all their reserves, internal and external, right to the breaking point, and this, they all knew, was when bad things usually started to happen.

Thus Kasjan declared that the next day, they would rest and organize their equipment for the final assault. Their "now or never" attempt would take place on October 19, when, as Vash wrote, "we'd go to the insuperable obstacle."

FIFTY-THREE

THE EXPLORERS SPENT OCTOBER 18 AS KASJAN had directed, sewing up tattered dry suits, collecting odds and ends of food, taking photos, telling stories, eating, and sleeping. They were all aware that they could well be on the verge of making history. After a huge supper, everyone went to bed at 9:25 P.M. Maddeningly, despite his exhaustion Vash was so excited that he could not sleep that night.

He got up early: 6:05 A.M. This was it, the sharp, final point to which everything had brought them. Vash brewed up a huge pot of "fragrant, bracing, morning coffee" to jump-start their day. At 9:00 A.M. they left camp; they arrived at the Large Fork ninety minutes later, bringing along heavy bags of rope and hardware to explore any breakthroughs.

In profile on a map, the new section of cave leading downward looks something like a decrepit stairway. Vash and Kyryl Gostev set about surveying one set of passages and pits, a task made more challenging for Vash, whose diarrhea required him to make increasingly frequent "trips to the hedgehogs," to use the quaint Ukrainian idiom for defecating in a cave. Vash's one-piece caving dry suit made things worse.

Surveying finished, the pair climbed back up to the Large Fork, rejoined the others, and then all descended to the place where Kasjan and Medvedeva had been working. Vash perceived at once that this section was both beautiful and promising. The limestone here reminded him of the delicate hues of raspberry and lemon sorbet, its surface studded with floral formations. The pit floors were strewn with stones that looked like elaborate white pastries. He began to sense that he might be close to achieving the kind of goal that, like participation in a great battle, divides a person's life into two distinct sections: before the event and after.

Here, too, the pitches were both wider and steeper—harbingers of depth. Presently, the entire team was gathered at the deepest point Kasjan and Medvedeva had reached. It was a small chamber, on the far side of which a steep pitch dropped off into darkness. It was eerily quiet. Kasjan was studying his altimeter intently. At some point during the last half hour, they had exceeded 2,000 meters—6,562 feet. This in itself was an epochal milestone, akin to the first ascent of an 8,000-meter peak. (That had been accomplished by the great French climber Maurice Herzog, on Annapurna, in 1950.)

As the others watched, Kasjan pulled a battered Snickers bar from one pocket, unwrapped the candy, munched some slowly, and shared the rest. *I will do my best, dear son.*

Then it was time. The next pitch would tell the tale. Kasjan adjusted his helmet lamps and slipped into the opening. Medvedeva followed, then Vash, then the others. They worked their way down three more vertical pitches separated by inclined passages and finally dropped into a triangular-shaped room, perhaps 15 feet on a side, with a flat, clay bottom and nondescript brown walls. Their light beams danced over floor and walls like frantic white spiders, but they could see no more windows, no more passages. This was it. The bottom of the world.

Everyone was quiet for a long time as the reality of where they were sank in. Vash, hard-pressed for words that adequately described his feelings of that moment, quoted the Russian poet Leonid Filatov:

Either the air is drunk or the hobgoblin is zealous.

Even in translation, that sentence suggests the rapture, positive at last, that Vash, Kasjan, and all of them were feeling. But they were not quite finished. In the center of the tricornered room's clay floor there was a circular, crater-like depression about 3 feet in diameter and 2 feet deep. Down at its bottom, the hole came to a point, like the drain in a sink (which it might well have been at one time), and in that spot was lodged a small white rock. Vash and the others leaned over the hole in Krubera's floor as Yury Kasjan used his altimeter to calculate the depth of that ultimate white rock. He straightened up, waited a moment, and announced, "Two thousand and eighty meters." The assembled team erupted with cheers.

These were educated, sophisticated scientists and explorers. They knew that they were experiencing one of the signal moments of history, the last link in a long, hallowed chain created by Peary at the North Pole, Amundsen at the South, Hillary and Norgay on Everest, Piccard and Walsh in the Challenger Deep, and many other, earlier greats who had paved the way for modern explorers. Kasjan and his people knew: they had just made the last great terrestrial discovery.

They hugged, cried, laughed, danced. It was, for Kasjan, as it would be for Klimchouk when he learned the news soon after, the brightest moment of a very eventful life. He was overcome by a rush of emotion—they all were. There were no dry eyes in that crowd. They shed tears, hugged, laughed, took pictures. Having made the discovery of a lifetime, they were feeling once-in-a-lifetime emotions that words struggle to capture. But think of the biggest joy you have ever felt and square it.

They were certainly feeling something else, as well. If you read the memoirs of great discoverers like those noted above, you know that a recurring theme is that, at their ultimate moments, those explorers were overjoyed but also *exhausted*. So were the members of Yury Kasjan's team, and the hardest part of their expedition remained undone.

No one wanted to leave, but they could not stay forever. Someone made a move toward the ascent passage.

Wait, Kasjan said. We have something still to do.

One final task remained. Discoverers in caves are privileged to name their new finds. It did not take long to reach consensus about the bottom of the world. They knew where they were, and how very long it had taken to get there, and how many other brave explorers had made possible this final act of a centuries-old drama.

They gave the place a name that, they hoped, captured all those things: Game Over.

AFTERWORD

On August 17, 2004, *The New York Times* reported that a team of Croatian cavers had "set a new benchmark that went largely unnoticed. They found the world's deepest hole." Interestingly, that article referred not to Krubera but to an unnamed pit that really was just a hole which plummeted 1,693 feet into a mountainside near the former Yugoslavian city of Zagreb. Later, the article acknowledged that the find was "not the deepest *cave* on Earth [emphasis added]. That title still belongs to the Krubera Cave in Abkhazia, which descends 5,130 feet (almost a mile)."

It was soon to descend a lot farther than that. Less than a week after the Croatians' find, Dmitry Fedotov and Denis Kurta worked through the Way to the Dream Meander, pushing Krubera's depth to 5,888 feet. Then, in October, Yury Kasjan descended to 7,072 feet, firmly establishing Krubera as the deepest cave and the last great terrestrial discovery. Krubera had one more surprise to spring. In August 2006, the Ukrainian cave diver Gennadiy Samokhin pushed its ultimate depth to 7,188 feet (2,191 meters) and its final length to more than 8 miles.

With his decades-long work on the Arabika Massif brought to fruition, Alexander Klimchouk has shifted his attention to another supercave, Aladaglar, in Turkey. While it won't surpass Krubera's depth, Aladaglar is still a true supercave, offering new subterranean challenges that draw Klimchouk and his teams.

Klimchouk divides his time between his work in the field, his academic responsibilities in Ukraine, and traveling the world. He is a sought-after speaker and presenter at international scientific conferences and is working on a book about his explorations.

Klimchouk remains estranged from his son Oleg; the two have not spoken for years.

BILL STONE CONTINUES TO PROBE the Cheve Cave system in Oaxaca, having led several expeditions there in the years after 2004. Cheve Cave's official depth is now 4,869 feet, appreciably less than Krubera's. In 2009, Stone led an ambitious but unsuccessful attempt to connect the lower cave called J2 with Cheve. He and his teammates spent the last nineteen days underground, diving one sump after another, ultimately mapping almost 2,000 feet of new passages. At the expedition's end, however, more than 3,000 feet still separated them from Cheve's deepest known point, far above. For the record, Stone still believes that Cheve has the potential to surpass Krubera.

On a separate track, Stone has moved a bit closer to fulfilling his astronaut dreams, building a NASA-funded interplanetary robot named Endurance. If all goes according to plan, sometime in the next decade Endurance will be flown to Jupiter's moon Europa, where it will search for water. Before that, though, Stone himself may go to our own moon. He has vowed, publicly, to establish the first commercial mining operation there by 2017.

ACKNOWLEDGMENTS

First, as always, thanks must go to my wife, Elizabeth Burke Tabor, without whose love and support this book would not have been written.

The remarkable microbiologist and caver Hazel Barton, a real-life Lara Croft, provided the idea that became this book. To thank her adequately for that is not possible, except to say that it was one of the greatest gifts I've ever received.

My intrepid and excellent agent, Ethan Ellenberg, helped shape the final concept and introduced me and it to Random House editor Jonathan Jao, about whom a special word is in order. Jonathan is a striking refutation of the assertion that "real editors" (i.e., in the Maxwell Perkins mode) are a thing of the past. Superb editing is every bit as much art and craft as writing, and Jonathan is inordinately gifted at both. It would take another book to chronicle his contributions to this one, but suffice to say that both I and Random House are infinitely lucky to have him on our side. If further proof of his worth were needed, it's provided by the fact that he is an avid Boston Red Sox fan.

Copy editors are truly the unsung heroes of book publishing, laboring anonymously to improve work for which they themselves rarely if ever receive credit. After one's agent and editor, however, a proficient copy editor is truly an author's best friend. It's impossible to overstate how much Random House's superb copy editor Bonnie Thompson improved this book, ensuring factual accuracy and polishing the prose far beyond my ability to do so alone.

Bonnie is not the only unsung hero at Random House, which does for me so many things I could never do myself. Without the help of marketing and publicity, a book is like an airplane on the tarmac without an engine. However sleek and beautiful the craft might be, without propulsion it will sit forever, a sadly grounded thing. Well, this book's "engines"—and they are powerful ones indeed—are marketing directors Sanyu Dillon and Avideh Bashirrad and publicity director Sally Marvin. For this book's stellar launch I can never thank them adequately, except to promise them, publicly, that I know how great is my debt to them.

Finally, even before Sanyu and Avideh and Sally can work their magic, they have to have a book. For that I have production editor Steve Messina to thank. Having been assisted by his counterparts in the magazine world, managing editors, I have a glimmer of an idea what an immense task it is to make real the vast and complicated idea we call a book. This is the place for me to honor Steve for birthing mine.

Bill Stone, one of the most remarkable men—and unquestionably the busiest one—I have ever met, graciously gave of his time and knowledge, patiently enduring visits, meetings, endless hours of interviews, and countless phone calls and emails, even while laboring in Antarctica. He supplied images, expedition logs, information, and hospitality and supported my research in every way possible. His trust, knowledge, and help were invaluable.

I owe a huge debt to *Beyond the Deep,* Bill Stone's own account of the fatal 1994 Huautla expedition, coauthored with Barbara am Ende and Monte Paulsen. An invaluable resource that provided records of contemporaneous conversations, thoughts, and experiences, it was the foundation of my research into that period of Bill Stone's life. I am immensely grateful to all three authors, both for their creation of this riveting book and for their permission to refer frequently to it. *Beyond the Deep* missed a well-deserved spot on the bestseller lists only through a bizarre series of misfortunes that are the stuff of every author's nightmares. My advice to those who enjoyed this book, as well as to all who relish fine writing about extraordinary people and achievements: read *Beyond the Deep.*

Carol Vesely recalled her and Bill Farr's discovery of Cheve and its early exploration. Barbara am Ende illuminated her and Bill Stone's explorations of Huautla. Andrea Hunter, Bart Hogan, John Kerr, Dave Kohuth, and Gregg Clemmer furthered my understanding of Cheve's exploration. I am indebted as well to Pat Kambesis, Nancy Pistole, Yvonne Droms, Diana Northup, Jeff Stolzer, Marcus Gary, Bill Torode, and Dave Bunnell.

Bill Mixon is perhaps the greatest living historian of Mexican caving; he and his world-famous private library were invaluable. John Schweyen recalled vivid details of his pioneering dive in Huautla. Bob Jefferys dug through long-buried archives to produce essential images and articles about the 1984 Peña Colorada expedition and his other cave exploits. The eminent British cave diver Rick Stanton shared his experiences on Stone-led expeditions, as did fellow U.K. cavers Robbie Warke and Paula Grgich. The outdoor writer Craig Vetter spoke candidly about his own times with Bill Stone. Geary, Sue, and Aspen Schindel provided bed and board in Texas and took me to the 2008 Texas Cavers Reunion, where I learned that nude mud wrestling is an essential part of cave-exploration training.

Alexander Klimchouk, like Bill Stone, supported my research in every way possible, submitting to personal visits, endless interviews, and everlasting interruptions by phone and email. In addition, he introduced me to eminent Russian and Ukrainian cavers. He pointed me toward a treasure trove of information available only in Russian, and then found the extraordinary translator Olga Rjazanova, who turned it all into readable English for me. Klimchouk spoke frankly not only about expeditionary caving but also about the personal toll it has exacted. His honesty helped put a very human face on what has been portrayed mostly as adrenaline-fueled adventure or dry science.

Yury Kasjan, second only to Klimchouk in fame and experience as a European supercaver, was similarly helpful. The eminent photographer Stephen Alvarez spoke candidly about his efforts chronicling Klimchouk-led Krubera expeditions, as well as American expeditions. Emil Vash shared his private expedition journals and images. Ekaterina Medvedeva provided valuable Krubera insights from a female perspective. Explorers Club fellow Chris Nicola, an important supercaver in his own right, helped me find and communicate with eastern European cavers. The Spanish caver Sergio García-Dils's articles about Krubera filled in many gaps. University of Florida historian Bogdan Onac's oral history interview with Alexander Klimchouk is the closest anyone has come to creating a biography of this remarkable man; it supplied many details of his life and career. French historian Pierre Olaf Schut shed light on Édouard A. Martel's life and career, as did the German historian Bernd Kliebhan.

Finally, the noted Australian cave explorer Alan Warild is the only person to have explored both Huautla and Krubera. His descriptions of the two, and his knowledge of the men who led their respective explorations, provided informative comparisons available from no other source.

Along its rocky road to completion, the manuscript received sensitive and essential reviews by Elizabeth Tabor, Damon Tabor, Jack Tabor, Sarah Ochs, Wallis Wheeler, Steven Butler, Tasha Wallis, and Sheila Bannister.

Without doubt I have omitted some who made signal contributions to the book. To them I apologize in advance, and offer here thanks from an eternally grateful author.

SUGGESTED READING

Cave exploration has generated nothing like the volume of literature produced by mountaineers and marine explorers, but there are still classics in the field. Those interested in the history of caving and speleology will enjoy Édouard A. Martel's books, in particular *Les Cévennes* and *Les Abîmes*. *La Côte d'Azur Russe* chronicles his visit to the Black Sea region and the Arabika Massif.

Martel's astonishing disciple Norbert Casteret was one of the most prolific writers in the history of exploration, publishing hundreds of articles and more than forty books. His classic *Ten Years under the Earth* (1933) remains an exhilarating and informative read to this day.

"Vertical Bill" Cuddington pioneered the single-rope technique (SRT) for vertical caving, now used as well by search-and-rescue (SAR) teams, rock climbers, and for industrial vertical work. The definitive biography is *Vertical Bill*, by David W. Hughes.

Sheck Exley was to cave diving as Bill Cuddington was to vertical caving. Before his tragic death in Mexico's El Zacatón cenote on April 6, 1994, he pioneered virtually every essential cave-diving technique and mentored hundreds, including Bill Stone. His *Basic Cave Diving: A Blueprint for Survival* and *Caverns Measureless to Man* are classics.

The best single source of information about Mexican supercave exploration is the Association of Mexican Cave Studies (www.amcs.org), operated by the legendary Texas caver Bill Mixon. Since 1975, the annual AMCS *Activities Newsletter* has contained articles about each year's significant Mexican caving expeditions. The quality of writing and photography is superb.

AMCS Bulletins, which appear periodically, are in-depth studies of single subjects, such as the one titled *Hydrogeology of the Sistema Huautla Karst Groundwater Basin* (2002).

The Russian geographer Alexander Kruber fostered speleology in his region of the world and exploration of the Arabika Massif. For those who read Russian (or are willing to pay for translations), his early articles, especially "The Voyage to Arabika" (1912), are rewarding.

Alexander Klimchouk has published scores of articles in scholarly and scientific journals, most of them in Russian. His fascinating account, in English, of breaking the 2,000-meter barrier in Krubera Cave, "In a Search for the Route to 2000 Meters Depth: The Deepest Cave in the World in the Arabika Massif, Western Caucasus," coauthored with Yury Kasjan, can be found in *Cavedigger*, issue 8, December 2003–February 2004.

Last but far from least are the countless articles published by Bill Stone, in both popular and scientific journals. His *Beyond the Deep*, written with Barbara am Ende and Monte Paulsen, is a detailed retelling of the ill-starred 1994 Huautla expedition.

NOTES

PART ONE: STONE

Chapter One

I learned the details of Chris Yeager's fatal accident in interviews and correspondence with Bill Stone, Tina Shirk (now Oliphant, but no relation to Matt Oliphant), and John Schweyen, all of whom were expedition members. Steve Knutson's official report of the accident, "Cueva Cheve, Oaxaca, Mexico, March 1 Aace—Caver fall, Equipment?," appeared in the December 1992 issue of *NSS News*. Louise D. Hose was also a member of the expedition. Her contemporaneous, detailed account of the fatality, included in her article "Exploration in the Sierra Juarez, Oaxaca: Cueva Cheve 1991–92," in the 1992 issue of the Association for Mexican Cave Studies's *AMCS Activities Newsletter*, was especially helpful. Another description of the accident appeared in "A History of Mexican Speleology to 1992," coauthored by Bill Stone and Terry Raines and published in the May 1997 issue of the *AMCS Activities Newsletter*. Other accounts appeared in *Rocky Mountain Caving*, the *Texas Caver*, and *Met Grotto News*.

Descriptions of caving vertical gear and its use came from my own experience. Technical expert William Storage's article "On Techniques and Safety," in the July 1993 *NSS News*, added to my knowledge of the many ways in which that gear can fail.

Chapter Two

In interviews and correspondence, Hazel Barton and Bill Stone described the Emily Davis Mobley accident and rescue in Lechuguilla. A segment of the Fox television series *Code 3* included detailed video of the rescue and interviews with Mobley herself. Other accounts appeared in *The New York Times* and the *Los Angeles Times*. My knowledge of cave-rescue systems and complications came from earlier interviews with cave-rescue legend Buddy Lane, who led the Mobley rescue effort.

Bill Stone, in interviews and correspondence, described his interaction with Chris Yeager's family and with the Mexican authorities.

Tina Oliphant told me about the 1992 recovery of Chris Yeager's body.

Comparisons between Bill Stone and the great mountaineer Reinhold Messner occurred to me early on in my research and, it turned out, had already appeared in several sources, including *Outside* magazine. Many articles and a number of books, by and about Messner, have described his climbing-related losses both on and off mountains.

Chapter Three

In interviews and correspondence, Carol Vesely described the details of her and Bill Farr's lives, as well as their discovery of Cheve Cave. Also helpful were the detailed accounts of many expeditions that she and Bill Farr wrote and collected in *Proyecto Cheve 1986–1993*. Their *AMCS Activities Newsletter* articles were also informative.

The terms "eight thousanders" and "8,000-meter peaks" are used to denote a unique category of mountains distinguished by size, elevation, and challenge. No similar term existed to give rare giants like Cheve and Krubera their due, so I coined "supercave" for that purpose.

The list of caving hazards, long but by no means inclusive, was drawn from my own experience and from the excellent work of William Storage, who has made a study of caving hazards and accidents. Especially helpful was his article "Using the Tools of Science and Industry to Build a Comprehensive Caving Safety Program," illustrated by Linda Heslop, which first appeared in the October 1991 issue of *NSS News*.

I experienced the memorable "uniquely alive smell" of deep caves myself.

Many books and articles have described native peoples' relationships with caves and their belief that caves are living things, including Benjamin Feinberg's outstanding *The Devil's Book of Culture: History, Mushrooms, and Caves in Southern Mexico.*

Bill Stone's book *Beyond the Deep*, coauthored with Barbara am Ende and Monte Paulsen, relates Stone's own encounters with Mazatec Indians who consider caves to be alive. The Mazatecs do not use the Spanish word *cueva* but instead refer to caves with the term *gui-jao*, which, as the noted Mazatec expert Renato García Dorantes says in the book, "represents something that is alive."

Angela M. H. Schuster's article "Rituals of the Modern Maya," in *Archaeology*, volume 50, number 4 (July–August 1997), sheds light on other contemporary Native Americans' beliefs about caves.

Chapter Four

Descriptions of Cheve's Entrance Chamber came from interviews and correspondence with Carol Vesely, Bill Stone, John Kerr, Andi Hunter, and others involved in exploring Cheve. *AMCS Activities Newsletter* articles contained similar accounts.

In the journal *American Antiquity*, volume 25, number 3 (January 1960), William R. Holland and Robert J. Weitlaner wrote about prehistoric human sacrifices performed by Cuicatecs whose descendants inhabit the Cheve region today. The modern Cuicatecs still perform blood sacrifices, but limit themselves (so it is assumed) to chickens and goats.

The veteran cave explorer Gary D. Storrick maintains a website, www.storrick.cnchost.com, that is arguably the world's single best collection of information about vertical caving and climbing devices, techniques, and history. John Cole's seminal invention of the rappel rack is acknowledged here and in many other caving histories.

Pierre Humblet, president of the International Mountaineering and Climbing Federation (UIAA), explained to me that rappelling, like success, has quite a few fathers. The French give credit for its creation to the Chamonix guide Jean Estéril Charlet. The Germans point to Hans Dulfer, for whom the Dulfer "seat" is named. The Italians cite Tita Piaz, with later improvements being made by Emilio Comici. Given that these cultures, the wellsprings of mountaineering, cannot agree among themselves, I thought it best to leave the attribution in the text general.

Chapter Five

Bill Stone provided details about his father, family life, introduction to caving, and early vertical work in interviews. Stone's sister, Judith Stone Jordan, graciously shared many memories of their family life in Ingomar.

Chapter Six

Bill Stone described in interviews and correspondence the creation of his ingenious, and mutually beneficial, deal with RPI's geology department. He also spoke about meeting his former wife, Pat Wiedeman, and their exploration activities together, including the 1982 ascent of Mount McKinley (now generally referred to by its Native American name, Denali). In addition to Stone's recollections, a number of accounts, replete with images, of Pat Wiedeman Stone's considerable contributions to supercave exploration exist in AMCS newsletters and other sources, including the United States Deep Caving Team's website, www.usdct.org. She was the only fully participating woman member of the 1984 Peña Colorada and the 1987 Wakulla Springs expeditions, which lasted 118 and 70 days, respectively.

Stone provided details of his Kirkwood days. The caving prodigy Jim Smith, who would go on to a legendary supercaving career of his own, told me about his world-record-setting descent into Gouffre de la Pierre Saint-Martin.

"Deep in such caves, explorers encounter watercourses big enough to satisfy the most avid whitewater kayakers": I witnessed such watercourses during cave descents in Tennessee, Alabama, and Georgia. Many other cavers I interviewed described such watercourses.

I learned about the history of Huautla Cave exploration in interviews and correspondence with Bill Stone, Bill Mixon, Barbara am Ende, Carol Vesely, Geary Schindel, and Bill Steele. *Huautla: Thirty Years in One of the World's Deepest Caves*, Steele's superb and comprehensive history, published in July 2009, was not available during my research, but it is a finely wrought recounting destined to become a mainstay of supercaving literature.

A note here about diving-related passages in the book, such as "That description really does not do the procedure justice, though, because you remain connected to your air tanks only by the regulator hose and mouthpiece": I'm a certified master diver myself, and have dived in the Atlantic, Pacific, Caribbean, Gulf of Mexico, St. Lawrence River, and Lake Champlain, as well as other lakes, quarries, and rivers. The depiction of the rigors of cave diving are drawn indirectly from my own experience, not in caves (I am not full-cave certified, and to enter caves without such training is tantamount to suicide) but in other "overhead environments," which include caverns, wrecks, dives in zero-visibility conditions, and dives in which decompression obligations prevent direct ascent to the surface.

In addition, I interviewed a number of accomplished cave divers, including Bill Stone, Barbara am Ende, John Schweyen, Rick Stanton, Bob Jefferys, Yury Kasjan, Ekaterina Medvedeva, Lisetta Wiese-Hansen, and Jim Parker.

Details about Bill Stone's harrowing San Agustín Sump dive came from interviews with Stone himself, as well as written accounts in his *Beyond the Deep* and articles in AMCS newsletters.

Chapter Seven

Bill Stone recounted details of his terrifying dive. The account in *Beyond the Deep* was also helpful.

Stone and co-leader Bob Jefferys told me about the 1984 Peña Colorada expedition. Stone's article "Peña Colorada" in the September 1984 *AMCS Activities Newsletter* was informative, as was Mark Minton's piece "Huautla Project," in the same issue.

Mark Minton's "Huautla Connection" in the December 1985 *AMCS Activities Newsletter* provided excellent updating on that exploration's progress. Stone's "Camping Beyond Sumps," also in that issue, offered a fascinating description of the evolving art and science of long-stay penetration of supercaves.

Descriptions of the joys of pitching proposals and desperately seeking funding were informed by interviews with Bill Stone and by my own experience.

Stone made the comment about Columbus during one of our interviews. He has offered the same opinion to other researchers, including John Tuttle in his article "Visionary Bill Stone Counting on Unmanned Vehicles," which appeared on December 12, 2007, on the Cyber Diver News Network website, www.cdnn.com.

Chapter Eight

In interviews and correspondence, Bill Stone and Bob Jefferys both recounted details of the 1984 Peña Colorada expedition, which they co-led. Bob Jefferys provided many unpublished images that, as the saying goes, were worth thousands of words. I found additional information and images in the expedition report on www.usdct.org.

Stone's article "The 1984 Peña Colorada Cave Expedition," in the *Explorers Journal*, volume 63, number 2 (June 1985), was very helpful. So was his "The Challenge of the Peña Colorada," *AMCS Activities Newsletter*, number 14 (September 1984).

Chapter Nine

Robert Forrest Burgess's classic work *The Cave Divers* provided much valuable information about cave diving's history and development. Two books by cave-diving pioneer and icon Sheck Exley, *Basic Cave Diving: A Blueprint for Survival* and *Caverns Measureless to Man*, were similarly helpful.

The fascinating and informative website www.dutchsubmarines.com described Cornelius Drebbel's contributions to the development of submarines and rebreathers. More information about Drebbel's inventions came from John H. Lienhard's article "Engines of Our Ingenuity No. 574: Cornelius Drebbel," on the website of the University of Houston's College of Engineering, www.uh.edu. The British Broadcasting Company's online history, www.bbc.co.uk/history, supplied additional information about Drebbel and his inventions, including a rendering in which Drebbel resembles a slimmed-down Wilford Brimley.

The story of the development of FRED came principally from interviews with Bill

Stone, descriptions in *Beyond the Deep*, and the U.S. Deep Caving Team's website, www.usdct.org. Also helpful was Stone's presentation "Deep/Underwater Cave Environments," given at the NASA Administrators Symposium "Risk and Exploration: Earth, Sea, and Stars." The symposium's report was edited and compiled by Steven J. Dick and Keith L. Cowing and is found online at www.spaceref.com. Finally, Stone's provocative "TED Talk" in March 2007 was also helpful. It is found online at www.ted.com.

Information about the navy's ill-fated EX-19 rebreather came from "EX 19 Performance Testing at 850 and 450 FSW (Feet of Seawater)," US Naval Experimental Diving Unit Technical Report NEDU-8-89.

Chapter Ten

I knew something about rigging and rebelays from my own caving experiences but learned vastly more from Bill Stone, Andi Hunter, Vickie Siegel, Bart Hogan, Robbie Warke, Paula Grgich, John Kerr, Gregg Clemmer, David Kohuth, Alexander Klimchouk, Yury Kasjan, and Emil Vash. A fascinating and informative account of rigging, written by the historian Bob Hoff, appeared in *Cave History Update*, December 15, 2003.

Chapter Eleven

Carol Vesely and Bill Stone described the March 1989 Cheve expedition in interviews and correspondence. Stone recounted the near tragedy of Meri Fish, as did several others I interviewed, including John Kerr and Andi Hunter.

Descriptions of living and working in darkness came from my own caving experience. The effects of prolonged darkness and isolation on the human system has been studied by a number of sources, including especially University of Texas social psychologist Sheryl Bishop, who has published a number of studies on the subject. "Evaluating Teams in Extreme Environments: Deep Caving, Polar and Desert Expeditions," presented at the 32nd International Conference on Environmental Systems in San Antonio, Texas, on July 1, 2002, was particularly helpful. I referred to several other publications by Bishop as well.

Possibly the most determined scientist to study darkness and isolation was the Italian sociologist Maurizio Montalbini, who died in September 2009, as I was writing these notes. In 1992–93, as part of a NASA-supported project to study the effects isolation and darkness might have on Mars-mission crews, Montalbini lived underground in a cave for 366 days—a record that still stands. He spent virtually the entire period in darkness. Among other noteworthy effects: his sleep-wake cycles doubled in length, his immune system's power dropped to zero, and he lost all sense of time. Emerging from his cave after 366 days, Montalbini was sure that only 219 days had passed. Results of this and other stays underground (Montalbini lived in caves for almost three years, total, during multiple studies) were published on his website, www.maurizioalbini.it and in *Advances in Space Biology and Medicine*, volume 3, December 1993.

"Cave darkness feels like water on a dive or air on a flight": I scuba dive and have flown both hang gliders and fixed-wing aircraft.

The descriptions of camping in caves came from my own experience as well as from interviews and correspondence with Bill Stone, Andi Hunter, Gregg Clemmer, John Kerr, David Kohuth, Barbara am Ende, Vickie Siegel, Alexander Klimchouk, Yury Kasjan, and Emil Vash.

Carol Vesely, Bill Stone, Andi Hunter, John Kerr, Gregg Clemmer, and others told me about the joys of negotiating breakdown mazes. I had done a few shorter ones myself.

Chapter Twelve

In interviews and correspondence, Bill Stone and Carol Vesely both described her amazing passage through the "impassable" breakdown.

Vesely and Bill Farr wrote about his penetration of Through the Looking Glass in *Proyecto Cheve 1986–1993.*

Bill Stone and Carol Vesely told me about the 1990 Cheve expeditions in interviews. In addition, Vesely and Farr wrote detailed accounts in *Proyecto Cheve 1986–1993.*

Jim Smith wrote about his 1990 dye-trace experiment in "Huautla Project," AMCS *Activities Newsletter,* January 1991.

Chapter Thirteen

John Schweyen told me about his sump dive in an interview.

Details of the fatal 1991 expedition came from interviews with Bill Stone, Tina Oliphant, John Schweyen, and other sources. Louise D. Hose's article "Exploration in the Sierra Juarez, Oaxaca: Cueva Cheve, 1991–92" in the *AMCS Activities Newsletter,* August 1992, was especially helpful. Hose was part of the 1991 expedition. Her lengthy, exhaustively researched piece includes interviews with many expedition members and a microscopic dissection of Chris Yeager's accident.

Descriptions of the 1993 expeditions appeared in "Proyecto Cheve Expedition 1993," AMCS *Activities Newsletter,* October 1993. To create this account, expedition member Mike Frazier compiled works by several authors, including himself, Nancy Pistole, Peter Bosted, and Peter Haberland.

Bill Stone described Brad Pecel's experience. It was also recounted in *Beyond the Deep.*

Chapter Fourteen

Descriptions of Rolf Adams's death came from interviews and correspondence with Bill Stone, Barbara am Ende, and Craig Vetter. I also relied on Vetter's articles "The Deep, Dark Dreams of Bill Stone" and "Bill Stone in the Abyss," which appeared in *Outside* magazine's November 1992 and November 1994 issues, respectively.

Bill Stone's tribute "Rolf Adams, 1965–1992" in the AMCS *Activities Newsletter,* August 1992, was also extremely helpful.

Having learned long ago that confessions, unlike revenge, are best served up warm, here's one: I could not bring myself to interrogate Pat Wiedeman about the breakup of her family. In my own experience, two divorces were among the most painful events of my life. Just imagining being asked probing questions about those agonies by a total stranger made my blood boil. The idea of me, a total stranger, interrogating Pat Wiedeman on the topic made my skin crawl, and I could not do it. Thus the narrative of those events in this chapter and others was based on facts Bill Stone volunteered to me and other interviewers, including Craig Vetter and Geoffrey Norman. Stone's descriptions of the dissolving marriage in *Beyond the Deep* were helpful as well. My intent was to indicate the chronology and the major cause of the divorce, which were all the story needed. My hope was to avoid opening old wounds and needlessly invading privacies. If I got things wrong, I apologize here to all concerned.

Chapter Fifteen

Descriptions and details of the first phase—that is, before Ian Rolland's death—of the 1994 Huautla expedition were drawn from interviews and correspondence with Bill Stone, Barbara am Ende, Craig Vetter, and other sources. That expedition was the primary focus of *Beyond the Deep*, which, as a result, was extremely valuable as a reference. So was Bill Stone's article about the expedition, "Huautla Cave Quest," in the September 1995 issue of *National Geographic*. I also referred to Anne Goodwin Sides and Hampton Sides's article "Journey Toward the Center of the Earth," in the August 28, 1994, *Washington Post Magazine*.

I was able to describe Ian Rolland's last dive in fine detail because of my own diving experience, the detailed recounting in *Beyond the Deep*, and, most of all, because Bill Stone and Barbara am Ende shared so generously with me memories that must have been exceedingly painful for both.

Chapter Sixteen

The details of Bill Stone and Barbara am Ende's meeting came from interviews and correspondence with both and from *Beyond the Deep*.

My descriptions of Barbara am Ende in 1994 are based on photographs taken of her at the time that are archived on www.usdct.org and on others that appeared in the September 1994 issue of *National Geographic*.

I learned about the animosity generated by am Ende's presence on the 1994 Huautla team from interviews with her, Bill Stone, Carol Vesely, Craig Vetter, and other sources. Craig Vetter's *Outside* magazine article "Bill Stone in the Abyss" contained numerous references to those bad feelings and directly quoted several disgruntled expedition members.

The relatively primitive state of supercaving communications was described for me in interviews with Bill Stone, John Kerr, and Alexander Klimchouk, among others.

In separate interviews, Stone and am Ende described the scene in Camp 3 after Kenny Broad delivered news of Ian Rolland's absence. *Beyond the Deep* also contained detailed depictions, as did Craig Vetter's *Outside* article.

Chapter Seventeen

The listing of hazards involved in recovering bodies from caves comes from my own training as a rescue diver, earlier interviews with cave-rescue leader Buddy Lane, more recent conversations with veteran cave divers Lisetta Wiese-Hansen and Jim Parker, and countless reports of such recoveries published by the National Speleological Society Cave Diving Section (NSSCDS) and the Divers Alert Network (DAN.) For those with the fortitude to view them, there are videos of cave-diving rescues gone bad available on YouTube. While some people might consider watching them the worst kind of voyeurism, they can serve an invaluable purpose: dissuading the unqualified from trying to dive in caves. It has always been true, and remains so, that the vast majority of cave-diving fatalities involve divers with little or no proper training.

Previous writings about the 1994 Huautla expedition gave short shrift—or none at all—to Bill Stone's recovery of Ian Rolland's body. Possibly this was because nondivers and noncavers could not appreciate the magnitude of the challenge involved. Possibly, also, it was because Stone, stung by accounts more focused on his personality than his achievements, had become increasingly wary of writers.

During our meetings, phone conversations, and email exchanges, Stone answered every question fully and without reservation, including those about the most painful experiences of his life, one of which was surely Ian Rolland's death. Without that kind of openness, the detailed account of his recovery of Rolland's body would not have been possible. The narrative in *Beyond the Deep* was also very informative.

Chapter Eighteen

An illuminating analysis of the complex tragedy that was the 1994 Huautla expedition was published in the *Journal of Human Performance in Extreme Environments,* volume 3, number 1. The authors were S. L. Bishop, P. A. Santy, and D. Faulk, senior researchers from the University of Texas's Department of Psychology. With Bill Stone's full approval, they interviewed and tested team members before and after the expedition. They also asked certain members to record their experiences in journals for later use by the researchers.

Bill Stone, Barbara am Ende, Carol Vesely, and Craig Vetter all described the mutinous and vituperative atmosphere that descended on the expedition after Rolland's death. Two very different versions of events, demonstrating once again the universality of *Rashomon*'s message, appeared in Vetter's 1994 *Outside* article and *Beyond the Deep,* not published until 2002.

Barbara am Ende's article "Off the Mainline: San Agustín Sump, Mexico" in the January–February 1995 issue of *Underwater Speleology,* a contemporaneous account of the 1994 Huautla expedition, provided much background for this and other Huautla-related chapters.

"I will tell you why this good man died": Several different versions of the elder Mazatec's specific words have been reported, but all convey the same essential message. This version came from Bill Stone, who was there and who was the old man's target.

Andi Hunter and Bill Stone both shared their recollections of the conversation between Stone and the young initiate.

Chapter Nineteen

"We beat the shit out of them": This quote appeared in Craig Vetter's 1994 *Outside* article.

Bill Stone and Barbara am Ende's lengthy article "The 1994 San Agustín Expedition" in the *AMCS Activities Newsletter,* May 1995, provided a wealth of information about all aspects of the expedition.

An account of the visit of the Mexican policemen appeared in *Beyond the Deep.* Bill Stone provided additional details.

Descriptions of Sloan's visit to the *curandero* came from interviews with Bill Stone, the *AMCS Activities Newsletter* article cited above, passages in *Beyond the Deep,* and Craig Vetter's *Outside* article.

Barbara am Ende told me about her discussions with Noel Sloan and their benign plot regarding diving beyond the sump with Stone.

To this day, no one knows with absolute certainty what killed Ian Rolland. But two reports written immediately after the death shed the most light. One was "Accident Analysis," a technical report coauthored by Bill Stone and Kenny Broad that appeared in the January/February 1995 issue of *Underwater Speleology*. The other was Barbara am Ende's article "Off the Mainline: San Agustín Sump, Mexico," in the same issue.

Chapter Twenty

Bill Stone and Barbara am Ende told me about their experiences diving past Sump 5. Excellent descriptions also appear in *Beyond the Deep* and in their article "The 1994 San Agustín Expedition" in the *AMCS Activities Newsletter*, May 1995. Am Ende's report in *Underwater Speleology*, noted earlier, was also helpful. Am Ende is an excellent cartographer, and I found her maps of Huautla invaluable in better understanding the character and conformation of this remarkable cave.

Chapters Twenty-one Through Twenty-three

The details of events in these chapters were drawn from interviews and correspondence with Bill Stone and Barbara am Ende, as well as from *Beyond the Deep*; their article "The 1994 San Agustín Expedition" in the May 1995 *AMCS Activities Newsletter*; Stone's article "Huautla Cave Quest" in the September 1995 *National Geographic* magazine; and am Ende's report "Off the Mainline: San Agustín Sump, Mexico" in the January/February 1995 issue of *Underwater Speleology*.

Chapter Twenty-four

The details of events in this chapter were drawn from interviews and correspondence with Bill Stone and Barbara am Ende, as well as from *Beyond the Deep*; their article "The 1994 San Agustín Expedition" in the May 1995 *AMCS Activities Newsletter*; Stone's article "Huautla Cave Quest" in the September 1995 *National Geographic* magazine; and am Ende's report "Off the Mainline: San Agustín Sump, Mexico" in the January/February 1995 issue of *Underwater Speleology*.

I have had the unnerving experience of squeezing through narrow, partly flooded passages in caves myself. But the most informative (and chilling) account I ever read of a flooded crack penetration was written by a caver named John Ackerman. In 1987, he set out to explore Tyson Spring Cave in Fillmore County, Minnesota. He described the harrowing experience in a December 23, 2006, report, "John Ackerman's Update on the Tyson Spring Cave," published on the Greek caving organization Zenas's website, www.zenas.gr/site/home/eng_detail.

> Eventually I came upon another sump but could see several inches of air space and so I decided to risk it once again. This almost proved to be a fatal error on my part because I actually became lost in this passage with my lips scraping against the ceiling. I meandered throughout the icy cold pitch-black passage, sniffing for a way out, any way out. My neck muscles were eventually so fatigued that they were almost unable to hold my lips to the ceiling. Finally I made the correct turn and popped out into the continuation of the huge cave passage. Hours later, after traveling almost two more miles through stupendous cave passages I turned around and made my long solo journey out. Along the way I was enthralled by the dynamics of the stream passage and understood my fate if it were to rain outside, causing the water to rise even one inch.

Chapter Twenty-five

The details of events in this chapter were drawn from interviews and correspondence with Bill Stone and Barbara am Ende, as well as from *Beyond the Deep*; their article "The 1994 San Agustín Expedition" in the May 1995 *AMCS Activities Newsletter;* Stone's article "Huautla Cave Quest" in the September 1995 *National Geographic* magazine; and am Ende's report "Off the Mainline: San Agustín Sump, Mexico" in the January/February 1995 issue of *Underwater Speleology.*

The quotes by and about Bill Stone in the first part of this chapter come from Craig Vetter's 1994 *Outside* article. Vetter confirmed, in an interview, that the magazine had not sent him to Mexico until rumors of bad happenings came drifting north.

Chapter Twenty-six

The quotes and information about Bill Stone's Wakulla Springs project came from interviews and correspondence with him, from Peter Symes's article about him in *X-ray Mag,* volume 15 (2007), and from Geoffrey Norman's article in the Summer 1999 *National Geographic Adventure.*

Chapter Twenty-seven

Andi Hunter told me about her introduction to Cheve, which included the festive birthday party arranged by Bill Stone.

John Kerr recounted his adventures on El Capitan and with the Cheve expedition.

I learned about the joys of digging in Cheve and Charco from interviews and correspondence with Bill Stone, Andi Hunter, John Kerr, Gregg Clemmer, David Kohuth, and Bart Hogan.

In separate interviews, Bill Stone and Andi Hunter described his "gardening" mishap.

Chapter Twenty-eight

John Kerr described his harrowing introduction to Cheve in interviews and correspondence.

Andi Hunter and other sources spoke about the *crinkle-crinkle* activities in camps.

R. D. Milhollin wrote about his 2003 Cheve descent in his excellent four-part article "Notes from Cheve," which appeared in the September, October, November, and December 2003 issues of the *Maverick Bull,* the monthly newsletter of the Maverick Grotto, the local caving club of Fort Worth and Tarrant County, Texas. The *Maverick Bull* is an outstanding example of the kind of caving literature produced regularly by grottoes all over the country. Milhollin's article "Caving in Sistema Cheve, Oaxaca," in the May 2004 *AMCS Activities Newsletter,* was also helpful. Yvonne Droms's report "Cheve 2003 Expedition News," in *NSS News,* August 2003, provided me with an overview of that expedition and was particularly helpful establishing the correct chronology of events.

Chapter Twenty-nine

Andi Hunter described her Rapture attack in an interview.

Bill Stone described the U.K. divers Rick Stanton and Jason Mallinson in an interview and in correspondence, during which he also shared his reservations about their minimalist rebreathers.

Rick Stanton's excellent article "Diving the Cheve Sumps," in the *AMCS Activities Newsletter*, May 2004, was also extremely helpful.

Chapter Thirty

In interviews, Rick Stanton described his predive ritual and the dive during which he and Jason Mallinson discovered Mad Man's Falls.

Information for this chapter was also obtained in interviews and correspondence with Bill Stone, Andi Hunter, Bart Hogan, Robbie Warke, and John Kerr.

PART TWO: KLIMCHOUK

Chapter Thirty-two

The description of Alexander Kabanikhin's accident was based on reports by Oleg Klimchouk, Denis Provalov, Julia Timoshevskayja, Bernard Tourte, and Sergio García-Dils. Alexander Klimchouk provided more details in interviews and correspondence.

Chapter Thirty-three

I learned about the 2005 helicopter crash from the report, with photographs, posted by "Stelios" on his blog, http://selas-voronya.blogspot.com/2005_04_01_archive.html. Alexander Klimchouk told me more in interviews and correspondence.

Alexander Klimchouk's biographical details came from interviews and correspondence with him. Also helpful was University of South Florida assistant professor Bogdan Onac's 2007 oral-history interview with Klimchouk, found online at http://kong.lib.usf.edu:8881.

Peter Grose's engrossing depiction of 1966 Kiev, "Kiev the Captivating," which appeared in the May 28, 1966, *New York Times*, included colorful details of that city in its pre-glasnost days.

Chapter Thirty-four

Interviews and correspondence with Alexander Klimchouk and Bill Stone informed this chapter, as did Bogdan Onac's oral-history interview with Klimchouk.

Chapter Thirty-five

Pierre-Olaf Schut's "E. A. Martel: The Traveller Who Almost Became an Academician," published in *Acta Carsologica*, volume 35, number 1 (2006), included many fascinating details about the pioneering French cave explorer's life and work. Dr. Schut also answered additional questions of mine about Martel.

The German historian Bernd Kliebhan has studied Martel's life as well. His website, www.kliebhan.de/spelhist/mar/mar-eng.htm, amounts to a mini-biography of Martel and provided much helpful information.

Alexander Klimchouk told me about his early Arabika explorations.

Chapter Thirty-six

Yury Kasjan told me about his early life and explorations. He and Alexander Klimchouk described Alexey Zhdanovich's breakthrough in Krubera.

Chapter Thirty-seven

Sergio García-Dils described his first Krubera expedition in "Nuevo Récord del Mundo de Profundidad: La Cima Krubera-Voronya," in the magazine *Subterránea,* issue 14 (2000). Yury Kasjan also described the expedition in an interview.

Alexander Klimchouk told me about the negotiations to bring CAVEX along on the 2000 expedition.

"In a Search for the Route to 2000 Meters Depth: The Deepest Cave in the World in the Arabika Massif, Western Caucasus," coauthored by Klimchouk and Kasjan, in *Cave-digger,* number 8, December 2004–February 2005, provided helpful information about the 2000 expeditions and much else related to the exploration of Krubera.

Chapter Thirty-eight

Alexander Klimchouk told me about the widening gap between himself and Oleg. He also described the media's treatment of CAVEX and the Ukr.S.A.

Chapter Thirty-nine

Alexander Klimchouk described the 2003 Krubera expedition in interviews and correspondence. Sergio García-Dils wrote a detailed account of the 2003 Krubera expedition, "Arábika—2003: En Busca del Premer –2000 del Planeta" in *Subterránea,* number 20 (2003). Additional accounts were written by Alexander Klimchouk, Yury Kasjan, and Nikoley Solovyev for www.speleogenesis.net. Yury Kasjan added further details in interviews and correspondence.

Chapter Forty

Details of the Kabanikhin rescue were drawn from accounts written by Sergio García-Dils, cited earlier, as well as from interviews with García-Dils and Alexander Klimchouk. Yvonne Droms's article "2003 Voronja (Krubera) Expedition News" in *NSS News,* November 2003, was also helpful.

PART THREE: GAME OVER

Chapters Forty-one Through Forty-five

Bill Stone, Bill Mixon, and Andi Hunter described Pedro Pérez's discovery of the Star Gorge Sink in interviews.

Information in this and following chapters about the 2004 USDCT Cueva Cheve expedition came from interviews and correspondence with Bill Stone, Andi Hunter, John Kerr, Gregg Clemmer, David Kohuth, and Bart Hogan. In addition, I was given access to the daily entries in the expedition's official Activity Log. Many team members wrote in the log, making it an excellent, *Rashomon*-like rendering of the expedition from multiple points of view.

Andi Hunter's article "Extreme Earth," which appeared in the *Explorers Journal,* Spring 2005, was as informative an account of Cheve exploration as it was a pleasure to read.

I also examined the full texts of all "Field Dispatches" sent daily to *National Geographic* magazine, one of the expedition's sponsors. Some, but not all, dispatches were later published as part of the article "Race to the Center of the Earth" in *National Geographic*'s April 2004 issue and on its website, www.NationalGeographic.com. I was able to review dispatches that were not published as well as unpublished portions of those that did find their way into print.

Finally, I had access to hundreds of images recorded by USDCT members and to extensive notes they had made about those images.

Chapter Forty-six

The "scooping booty" incident referred to in this chapter was described in interviews with Bill Stone, Andi Hunter, John Kerr, and Bart Hogan.

Chapter Forty-eight

In interviews and correspondence, Alexander Klimchouk, Yury Kasjan, Emil Vash, Ekaterina Medvedeva, Stephen Alvarez, and Marcus Taylor described Krubera's forbidding character. In addition, all recounted details of the August 2004 expedition.

Other helpful descriptions of Krubera appeared on the websites www.cavex.ru and www.selas-voronya.com. Images provided by Alexander Klimchouk, Yury Kasjan, Ekaterina Medvedeva, and Emil Vash, accompanied by notes, augmented those descriptions. The English photographer and videographer Marcus Taylor gave me illuminating still images and video footage as well. Finally, photos by the noted American photographers Stephen Alvarez and Alan Cressler, published along with Alexander Klimchouk's article "Call of the Abyss—the World's Deepest Cave," *National Geographic* magazine, May 2005, were helpful.

Chapters Forty-nine Through Fifty-three

Yury Kasjan and Ekaterina Medvedeva shared with me the details of their experiences during the October 2004 Krubera expedition. Also helpful was Alexander Klimchouk's article in the May 2005 issue of *National Geographic*, cited above.

I am especially indebted to Emil Vash, who gave me full access to his private 2004 Krubera expedition journal.

INDEX

PHOTO © STEPHEN BUTLER

JAMES M. TABOR's last book was the international award winner *Forever on the Mountain*, published in 2007. He was the co-creator and executive producer of the 2007 History Channel special about the exploration of Naj Tunich Cave, *Journey to the Center of the World*. He was also the writer and on-camera host of the popular national PBS adventure series *The Great Outdoors*. Tabor's writing has appeared in *Time*, *Smithsonian*, *U.S. News & World Report*, *The Wall Street Journal*, *The Washington Post*, *Barron's*, and many other national publications. He is a former contributing editor to *Outside* magazine and *Ski Magazine* and is at work on a thriller to be published by Ballantine Books in 2012.